Springer Theses

Recognizing Outstanding Ph.D. Research

Aims and Scope

The series "Springer Theses" brings together a selection of the very best Ph.D. theses from around the world and across the physical sciences. Nominated and endorsed by two recognized specialists, each published volume has been selected for its scientific excellence and the high impact of its contents for the pertinent field of research. For greater accessibility to non-specialists, the published versions include an extended introduction, as well as a foreword by the student's supervisor explaining the special relevance of the work for the field. As a whole, the series will provide a valuable resource both for newcomers to the research fields described, and for other scientists seeking detailed background information on special questions. Finally, it provides an accredited documentation of the valuable contributions made by today's younger generation of scientists.

Theses may be nominated for publication in this series by heads of department at internationally leading universities or institutes and should fulfill all of the following criteria

- They must be written in good English.
- The topic should fall within the confines of Chemistry, Physics, Earth Sciences, Engineering and related interdisciplinary fields such as Materials, Nanoscience, Chemical Engineering, Complex Systems and Biophysics.
- The work reported in the thesis must represent a significant scientific advance.
- If the thesis includes previously published material, permission to reproduce this must be gained from the respective copyright holder (a maximum 30% of the thesis should be a verbatim reproduction from the author's previous publications).
- They must have been examined and passed during the 12 months prior to nomination.
- Each thesis should include a foreword by the supervisor outlining the significance of its content.
- The theses should have a clearly defined structure including an introduction accessible to new PhD students and scientists not expert in the relevant field.

Indexed by zbMATH.

More information about this series at https://link.springer.com/bookseries/8790

Yancai Yao

Controllable Synthesis and Atomic Scale Regulation of Noble Metal Catalysts

Doctoral Thesis accepted by
University of Science and Technology of China, Hefei, China

Springer

Author
Dr. Yancai Yao
Department of Chemistry
University of Science and Technology
of China
Hefei, China

Supervisor
Prof. Yuen Wu
School of Chemistry and Materials Science
University of Science and Technology
of China (USTC)
Hefei, China

ISSN 2190-5053 ISSN 2190-5061 (electronic)
Springer Theses
ISBN 978-981-19-0204-8 ISBN 978-981-19-0205-5 (eBook)
https://doi.org/10.1007/978-981-19-0205-5

This Springer imprint is published by the registered company Springer Nature Singapore Pte Ltd.
The registered company address is: 152 Beach Road, #21-01/04 Gateway East, Singapore 189721, Singapore

Supervisor's Foreword

Improving the atom utilization efficiency, particularly for high-cost noble metals materials, in various chemical transformations, is of paramount significance for sustainable chemistry. A challenge in implementing this pioneering goal is to develop fundamental strategies that can well engineer noble metal nanocrystals at the atomic scale. In this thesis, Yao mainly focuses on developing the controllable methods including epitaxial growth, electrochemical leaching and acid etching that can finely tune the atomic structure of noble metal nanocrystals. Also, she demonstrated the promising applications of atomically level noble metal materials in heterogeneous catalysis field. I believe this thesis can offer guidance for researcher to deliberately design atomically level noble metal materials.

Hefei, China
December 2021

Prof. Yuen Wu

Abstract

Metal nanocrystals have a wide range of applications in catalysis, medical, sensing and energy storage. It is well known that the size, composition, morphology and crystal structure of metal nanocrystals can largely affect their performance. Therefore, the preparation of metal nanocrystals with controlled crystal structure allows us to explore their structure–properties relation. Generally, modulating the crystallographic structure of a bulk metal requires high temperature and pressure, which is not favorable to practical applications. Actually, surface stress of nanocrystals plays an important role in controlling the crystal phase structure. Scientists have put a lot of efforts to modulating the crystal phase structure of nanocrystals with a mild method, yet is still challenging.

By developing efficient and durable electrolyzers for water splitting, the intermittent electrical energy generated from renewable wind and solar energy can largely be converted into fuels. Water electrolysis using a polymer electrolyte membrane (PEM), based on proton transfer, has been demonstrated to effectively mitigate the drawbacks of an alkaline environment, including the crossover of product gases, limited current density and low operating pressure. Unfortunately, the sluggish oxygen evolution reaction (OER) and intense degradation of the catalysts in low pH and the strong oxidative environment impede the widespread adoption of practical electrolyzers. Compared with Ir-based systems with better dissolution resistance, Ru has more abundant reserves and been evaluated to be a more active OER catalyst due to its lower overpotential. However, commercial RuO_2 is easily oxidized to RuO_4 under strong acid and oxidative environment, resulting in its deactivation, which is arisen from the lattice oxygen participating in the evolution of oxygen during OER. Therefore, developing a highly efficient and durable Ru-based material for oxygen evolution is urgently desired, yet challenging.

The excessive exploitation of fossil fuels leads to the extensive release of greenhouse gases, such as carbon dioxide (CO_2), nitrous oxide (N_2O), hydrofluorocarbons (HFCs) and sulfur hexafluoride (SF_6), which pose a risk to human health and result in polluted environment. Therefore, developing alternative renewable energy to fossil fuels is urgently required. Hydrogen as a clean energy carrier in fuel cells is industrially produced from carbon feedstocks via various processes, which are environmentally unfriendly because of the release of CO_2. The electrochemical hydrogen evolution reaction (HER) in water electrolysis is an alternative way to produce H_2 without the release of CO_2. This process is still largely limited to the use of precious metals such as Pt, especially for industrial requirements. This largely increases the cost of electrolyzers and prevents their large-scale applications. Therefore, constructing Pt catalysts with low loading and efficient performance for HER is highly required.

We report an epitaxial growth-mediated method to grow face-centered cubic (FCC) Ru, which is thermodynamically unfavorable in the bulk form, on the surface of Pd–Cu alloy. Induced by the galvanic replacement between Ru and Pd–Cu alloy, a shape transformation from a Pd–Cu@Ru core–shell to a yolk–shell structure was observed during the epitaxial growth. The successful coating of the unconventional crystallographic structure is critically dependent on the moderate lattice mismatch between the FCC Ru overlayer and $PdCu_3$ alloy substrate. Further, both FCC and hexagonal close packed (hcp) Ru can be selectively grown through varying the lattice spacing of the Pd–Cu substrate. The presented findings provide a new synthetic pathway to control the crystallographic structure of metal nanomaterials.

We have synthesized a series of $PtCu_x/Pt_{skin}$ core–shell structures with atomically dispersed Ru_1 and have unraveled their mechanism of formation, oxidation resistance and the origin of the enhanced OER catalysis by exhaustively examining their structures, coordination environments and oxidation states. Through sequential acid etching and electrochemical leaching, the structure of $PtCu_x$ alloys can be varied (to give $PtCu_3$, PtCu and Pt_3Cu), which modulates effectively the OER activity catalyzed by the Ru_1. The best catalyst, Ru_1–Pt_3Cu, delivers 220 mV overpotential to achieve a current density of 10 mA cm^{-2} for acidic OER, with ten times longer lifetime over commercial RuO_2. We found that there is a volcano-type relation between the OER activity and the lattice constant. We argue that the compressive strain of the Pt_{skin} shell effectively engineers the electronic structure and redox behavior of single atomic Ru anchored at the corner or step sites of the Pt-rich shell, with optimized binding of oxygen intermediates and better resistance to over-oxidation and dissolution.

Submonolayer Pt was controllably deposited on an intermetallic Pd_3Pb nanoplate (AL-Pt/Pd_3Pb). The atomic efficiency and electronic structure of the active surface Pt layer were largely optimized, greatly enhancing the acidic HER. AL-Pt/Pd_3Pb exhibits an outstanding HER activity with an overpotential of only 13.8 mV at 10 mA/cm^2 and a high mass activity of 7834 A/g_{Pd+Pt} at −0.05 V, both largely surpassing those of commercial Pt/C (30 mV, 1486 A/g_{Pt}). In addition, AL-Pt/Pd_3Pb shows excellent stability and robustness. Theoretical calculations show that the improved activity is mainly derived from the charge transfer from Pd_3Pb to Pt, resulting in a strong electrostatic interaction that can stabilize the transition state and lower the barrier.

Keywords Noble metal nanocrystal · Controlled synthesis · Atomic scale · Electrocatalysts

Acknowledgements

Time flies. I am coming to the end of my Ph.D. studies, which are happy memories in my life. I will never forget this impressive journey, during which I would like to appreciate my teachers and friends who always support and help me to accomplish my Ph.D. studies.

Firstly, I would like to thank Prof. Yadong Li, who gave me an opportunity to study in the Center for Advanced Nanocatalysis (CAN) of USTC. It is my great honor to receive your guidance in my Ph.D. studies process. Also, your rigorous and diligent attitude toward academic research has deeply influenced, inspired me and promoted me to further explore scientific research.

Secondly, I would like to thank Prof. Yuen Wu. He had devoted tremendous efforts to help me finish my Ph.D. studies involving the selection of research field, experimental design, experiments executions and conclusion analysis. Professor Wu gave me great encouragement when I encountered bottlenecks in the implementation of the project and helped me to analyze and solve questions. I still remembered one subsentence you said "As long as you can do one thing persistently, you can do well," which made a great impact on me. Thanks to Prof. Wu for all he has done for me.

Thirdly, I would like to thank other teachers who helped me a lot with my studies. Thanks to Prof. Zhaoxiang Deng and Prof. Hong Xun from CAN in providing help for studies guidance. Thanks to Prof. Weixue Li and Sulei Hu from the department of physics chemical of USTC in providing help for theoretical calculation. Thanks to Prof. Dongsheng He from Southern University of Science and Technology (SUST) in providing help for TEM characterizations and analysis. Thanks to Prof. Xusheng Zheng and Qian Xu from Hefei National Synchrotron Radiation Laboratory in providing help for XPS measurements and analyses.

Also, I would like to thank my brothers and sisters in my group: Daping He, Peiqun Yin, Yunteng Qu, Jing Wang, Zhijun Li, Tieyun Zhao, Jingjie Ge, Yiqi Luo, Jian Yang, Xin Wang, Geng Wu, Huang Zhou, Yafei Zhao, Xiaoqian Wang, Changming Zhao, Chao Zhao, Mengzhao Zhu, Tingting Chao, Yanmin Hu, Zhenggang Xue, Tongwei Yuan, Wenyu Wang, Min Chen, Fangyao Zhou, Rongbo Sun, Benjin Jin, Yan Zhang, Xianyan Zhao, Peigen Liu, Lin Tian, Qi Xu, Weichen Wei, Haiyuan Lu and so on. Our group is a cohesive team with dream and passion. Wish you all better and better.

In particular, I would like to thank a lot for my parents and brothers. Your support and encouragement have carried me along and given me the courage to move forward. Thanks to my husband Shan Qin for his company and support.

Finally, I would like to thank all the experts and scholars who review my doctoral dissertation. Thanks a lot for your efforts.

May 2019 Yancaı Yao

Contents

1 Introduction 1
1.1 Introduction of Noble Metal Nanocrystals 1
1.2 Methods for the Preparation of Noble Metal Nanocrystals 2
1.2.1 Solvothermal and Hydrothermal Method 2
1.2.2 Coprecipitation Method 2
1.2.3 Sol–Gel Method 5
1.2.4 Ultrasonic Method 6
1.2.5 Microwave Method 7
1.3 Characterizations of Noble-Metal Nanocrystal 9
1.4 The Application of Noble Metal Nanocrystalline 10
1.4.1 The Application of Noble Metal Nanocrystals in Organic Catalysis 10
1.4.2 The Application of Noble Metal Nanocrystals in Electrocatalysis 12
1.4.3 Application of Noble Metal Nanocrystals in Surface-Enhanced Raman Scattering (SERS) 14
1.5 Electrocatalytic Water Splitting Technology 15
1.5.1 Introduction of Electrocatalytic Water Splitting Technology 15
1.5.2 Oxygen Evolution Reaction (OER) 16
1.5.3 Electrocatalytic Hydrogen Evolution Reaction (HER) 20
1.6 The Research Content and Purpose 23
References 24

2 Modulating FCC and HCP Ruthenium on the Surface of Palladium–Copper Alloy Through Tunable Lattice Mismatch 33
2.1 Introduction 33
2.2 Experimental Section 34
2.2.1 Materials 34
2.2.2 Methods 35
2.2.3 Characterizations 35

2.2.4 Electrochemical Measurements 36
2.2.5 Organic Catalysis 36
2.3 Results and Discussion 37
2.4 Conclusions 52
References 52

3 Engineering the Electronic Structure of Single Atom Ru Sites via Compressive Strain Boosts Acidic Water Oxidation Electrocatalysis 55
3.1 Introduction 55
3.2 Experimental Section 57
3.2.1 Materials 57
3.2.2 Methods 58
3.2.3 Characterizations 58
3.2.4 Electrochemical Measurements 60
3.3 Results and Discussion 62
3.4 Conclusion 83
References 89

4 Engineering the Electronic Structure of Submonolayer Pt on Intermetallic Pd_3Pb via Charge Transfer Boosts the Hydrogen Evolution Reaction 93
4.1 Introduction 93
4.2 Experiments 94
4.2.1 Materials 94
4.2.2 Methods 94
4.2.3 Characterizations 95
4.2.4 Electrochemical Measurements 95
4.3 Results and Discussion 96
4.4 Conclusion 107
References 114

5 Conclusion and Perspective 117

Author Biography 119

Chapter 1
Introduction

1.1 Introduction of Noble Metal Nanocrystals

Nanomaterials are defined as nanoscale materials with a size range of 1–100 nm in the three-dimensional space [1]. They feature a single-domain crystal without complex grain boundaries. As a bridge between atoms and bulk materials, nanocrystals received a booming research interest due to their unique properties. Therefore, controllable synthesis of nanocrystals is essential for the development of science and technology.

Noble metal nanomaterials are defined as nanocrystalline materials containing noble metal elements, such as gold (Au), silver (Ag) and platinum group metals (e.g., ruthenium (Ru), rhodium (Rh), osmium (Os), palladium (Pd), iridium (Ir) and platinum (Pt)). Noble metals are chemically stable in the earth and they are widely used in the fields of optics [2], biosensing [3–5], catalysis [6, 7], and electronics [8, 9]. Thus, the preparation, characterization and application of noble metal nanocrystals have attracted extensive attention.

The properties of noble metal nanocrystals are largely determined by physical parameters such as size, morphology, composition and structure (e.g., hollow or solid). Thus, the properties of noble metal nanocrystals can be well modulated by fine-tuning these physical parameters. For example, Pt nanocrystals exhibit discrepant catalytic activity in acidic oxygen reduction reaction (ORR) with different facets exposure [10]. Recently, researchers have made great progress in manipulating materials at the nanoscale, enabling us to better understand the geometric, crystal and electronic effects of nanocrystals and further discover the potential applications of nanocrystals in new field. Therefore, the preparation, characterization and application of noble metal nanocrystals are discussed below.

Y. Yao, *Controllable Synthesis and Atomic Scale Regulation of Noble Metal Catalysts*, Springer Theses, https://doi.org/10.1007/978-981-19-0205-5_1

1.2 Methods for the Preparation of Noble Metal Nanocrystals

The advancement of nanoscience and nanotechnology largely depends on the controllable synthesis of nanomaterials with different morphologies and sizes to a large extent. Regarding that the properties of metal nanocrystals are closely related to their size, morphology, composition, structure and crystallinity, researchers have developed various methods to manipulate the nanomaterials in the synthesis process to meet industrial requirements. At present, researchers have prepared nanocrystals with different morphologies, sizes, compositions and functions through physical, chemical and biological methods. In this section, we mainly focus on the preparation methods of nanocrystals.

1.2.1 Solvothermal and Hydrothermal Method

Solvothermal and hydrothermal processes occur in a closed system (for example, an autoclave) with high temperature and pressure. During the temperature elevating, the pressure in the autoclave increases, causing the solvents to exceed their boiling point. The one using water as the solvent is called the hydrothermal process, and the one using non-aqueous solvents is referred as the solvothermal process [11, 12]. At present, synthesizing metal nanocrystals by solvothermal and hydrothermal method is widely developed.

In 2012, our group synthesized a series of shape-controlled water-soluble Pt, Pt_xNi_{1-x} through a solvothermal method [13]. Briefly, platinum and nickel precursors are uniformly dispersed in the solvent by ultrasonic, and subsequently the mixture is transferred to an autoclave for heating. Pt-Ni nanocrystals with different morphologies were prepared following the above procedure by changing capping agents (Fig. 1.1).

In addition, Yang et al. prepared high-yield Ag nanocrystals with different morphologies (triangle, hexagon, cube, nanorods and polyhedron) through a one-step solvothermal reduction [14]. First, N,N-dimethylformamide (DMF) solvent, PVP (weak reducing agent, stabilizer and capping agent) and silver nitrate ($AgNO_3$) were mixed, and then the mixed solution was transferred to the autoclave with high pressure and high temperature treatment. By adjusting the amount of $AgNO_3$ and PVP, Ag nanoparticles with different morphologies were synthesized (Fig. 1.2).

1.2.2 Coprecipitation Method

Most of the earliest research focus on the preparation of nanocrystals by the coprecipitation of slight soluble products in an aqueous solution, followed by thermal

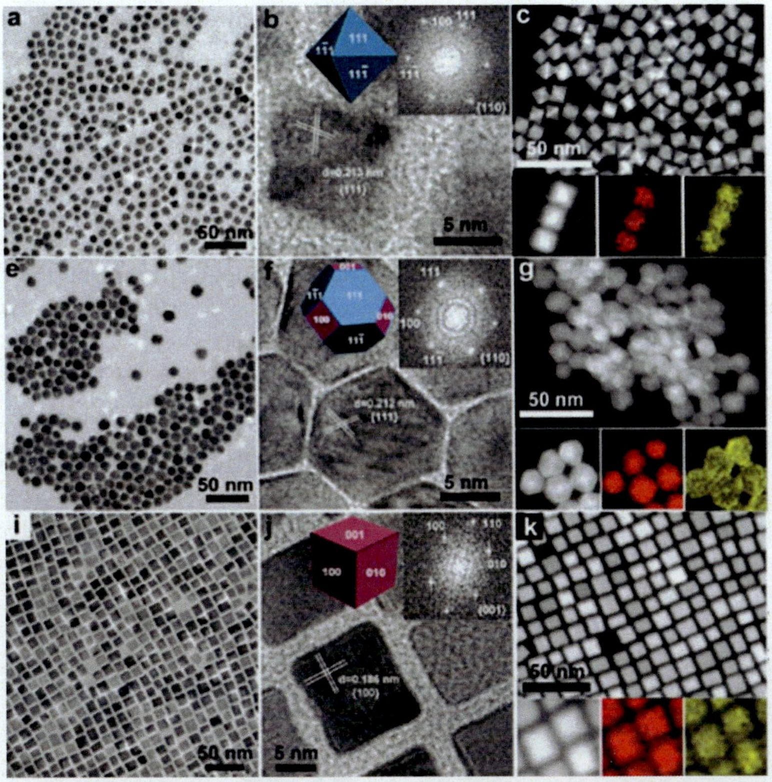

Fig. 1.1 Pt-Ni nanocrystals with different morphologies prepared by solvothermal method [13]

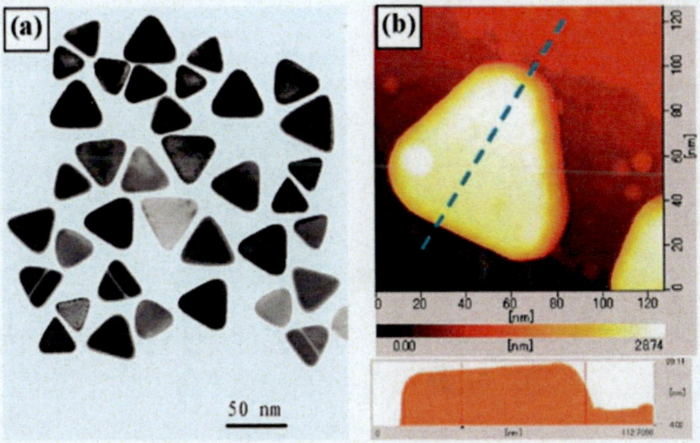

Fig. 1.2 Ag nanoparticles prepared by solvothermal method [14]

treatment to obtain nanoparticles. The coprecipitation reaction involves spontaneous nucleation, growth, crystallization/condensation and other processes. The formation of metal precipitation in aqueous or non-aqueous solution first requires the reduction of metal ions. Most commonly used of the reductants are various, such as H_2, ABH_4 (A = alkali metal), hydrazine hydrate ($N_2H_4{\cdot}H_2O$) and hydrazine hydrochloride ($N_2H_4{\cdot}2HCl$).

The coprecipitation process tends to involving the following four characteristics: (1) The product is usually slight soluble species in a highly saturated solution. (2) Nucleation is a key step in the coprecipitation process, and this process forms lots of small crystal nuclei. (3) Secondary process, such as Ostwald ripening and aggregating, largely affect particle size, morphology and thus performance. (4) Coprecipitation induced by a saturated solution often originates from a chemical reaction. Likewise, the adding rate of reactants and stirring speed also affect the size, morphology and size distribution of the product.

Coprecipitation method is also applied to the preparation of noble metal materials. For example, Tan et al. successfully prepared noble metal nanocrystals such as Au, Pt, and Pd with potassium bitartrate as the reductant [15] (Fig. 1.3). Similarly, the monodispersed Ag nanoparticles with the size of about 3.3 nm were prepared by employing borohydride as the reductant [16].

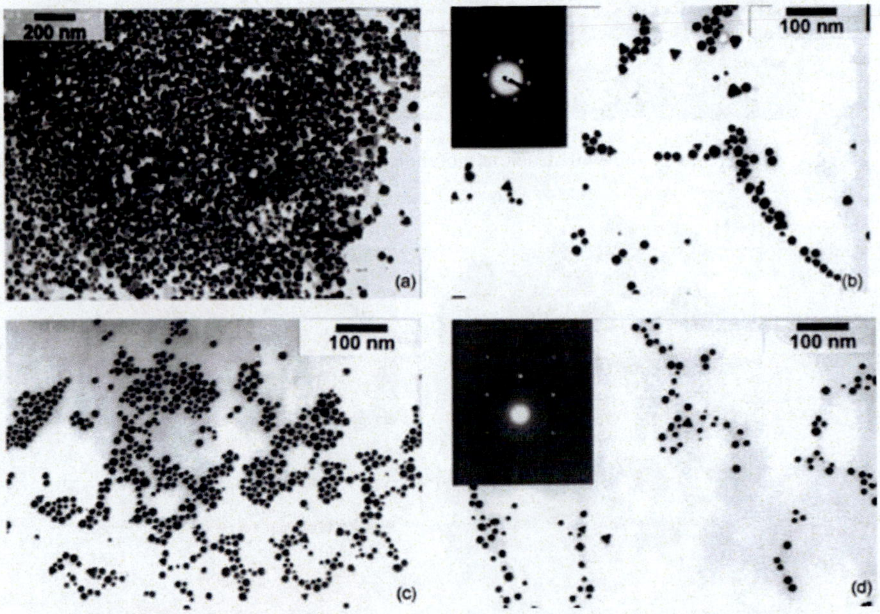

Fig. 1.3 TEM micrographs of Au nanoparticles synthesized using coprecipitation method [15]

1.2.3 Sol–Gel Method

The sol–gel method has attracted much attention with the development of the controlled preparation of noble metal nanomaterials. The synthesis of noble metal nanocrystals by sol–gel method involves chemical, photochemical or electrochemical reduction of metal salts or ultrasonic chemical/thermal decomposition of metal compounds, which occurs usually in aqueous solution or organic solvent with a variety of additives (such as surfactants, ligands and polymers, etc.). The commonly used sol–gel methods for preparing noble metal nanomaterials include seed-induced growth [17–20], high temperature reduction [21–24], template induction [25–28], electrochemical synthesis [29–32], photochemical synthesis [33–36]. In the following part, seed-induced growth and electrochemical synthesis are mainly introduced.

Seed-induced growth is widely utilized in sol–gel chemistry to synthesize noble metal nanocrystals. In the growth process, small metal particles as crystal seed were obtained by reducing metal ions with the appropriate reductants. The crystal seed is then added to the growth solution containing the same or different metal ions. With the assistance of additives like ligands, stabilizers and reductants, the metal ions in the growth solution are reduced on the surface of the crystal seed through heterogeneous nucleation. By this seed-induced sol–gel method, researchers prepared lots of nanocrystals with various morphologies (Fig. 1.4), such as linear, planar, 3D and dendritic, etc. [37–40].

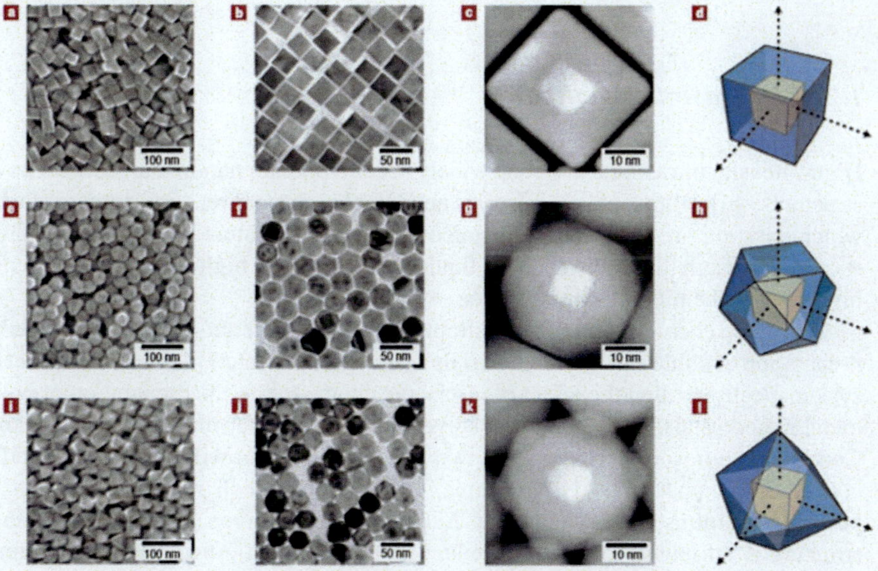

Fig. 1.4 Binary metal nanocrystals through seed-induced growth [38]

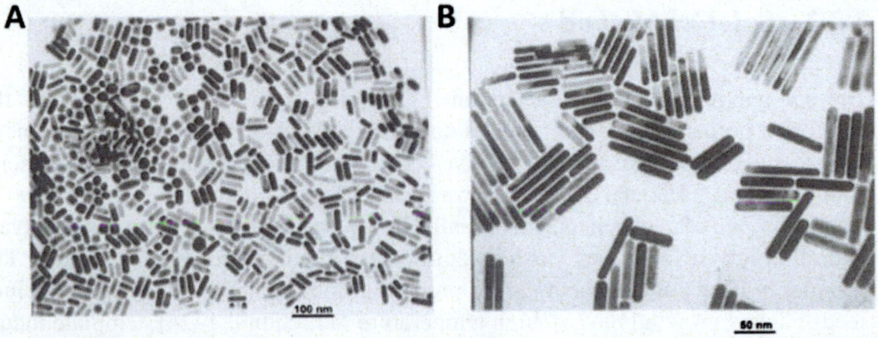

Fig. 1.5 Nanocrystals synthesized by sol–gel method [43]

Discharge technology is also one of the classical sol–gel methods. The preparation of nanoparticles via electrochemical method is a chemical reaction with applied potentials in the electrolyte. The size and morphology of nanoparticles can be controlled by changing electrodeposition parameters, including potential, temperature, deposition time, surfactant and polymer [41, 42]. Yu et al. reported that nanorods with different aspect ratios and facets exposure were prepared by electrochemical method [43, 44] (Fig. 1.5). Zhu et al. also prepared triangular Au nanosheets by electrochemical method [45]. Consequently, bottom-up sol–gel chemical synthesis has become a powerful tool to manipulate noble metal nanocrystals with controllable morphology.

1.2.4 Ultrasonic Method

High-intensity ultrasound has been widely used to prepare nanomaterials with novel structures [46]. Ultrasonic chemistry is an extreme transient reaction process, during which hot spots are formed accompanied with the temperatures to 5000 K and pressures to 1000 atm [47, 48]. When the liquid is irradiated by high-intensity ultrasound, high-energy chemical reaction occurs.

Ultrasonic chemistry provides a simple and efficient method for the controllable preparation of noble metal nanomaterials (Au, Ag, Pt, Pd, etc.) [49–56]. Grieser et al. systematically studied the effect of ultrasonic chemistry on the preparation of noble metal nanocrystals, suggesting that solvent/surfactant determined the particle size. Continuous ultrasonic reduction of two different Au ions will produce bimetallic core–shell nanoparticles. For example, Au@Pd core–shell nanoparticles can be prepared by ultrasonic reduction of Au(III) and Pd(II) ions [57]. The core–shell structure is attributed to different reduction potentials of Au(III) and Pd(II) ions. Besides, noble metal nanoparticles with core–shell structure such as Au/Ag and Pt/Ru can also be synthesized by the similar method [58].

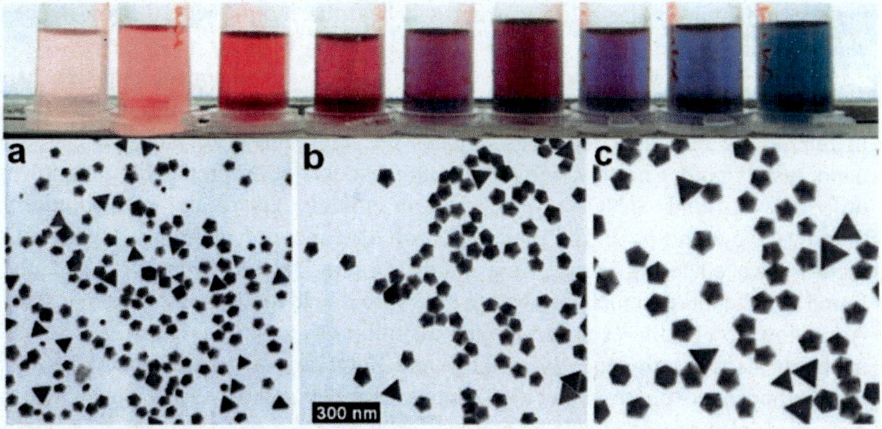

Fig. 1.6 TEM images of decahedral gold nanoparticles synthesized by ultrasonic method [60]

In addition, the ultrasonic chemistry can also be unitized to prepare non-spherical nanoparticles. For example, Au nanorods were obtained by ultrasonic reduction of $HAuCl_4$ in $AgNO_3$, CTAB and AA solutions [59]. It was demonstrated that the solution pH affected the average aspect ratio of Au nanorods. The average aspect ratio of Au nanorods decreases with the pH increasing. When $HAuCl_4$ was ultrasonically reduced with PVP as a stabilizer in Au seed solution, monodisperse Au decahedral would be obtained [60] (Fig. 1.6).

Ag nanosheets can also be prepared by ultrasonic-assisted Ostwald ripening process with Ag nanoparticles as seeds [61]. Using nanosheets as crystal seeds, annular nanostructures can be synthesized by ultrasonic-assisted method. In addition, Ag nanowires and nanorods can also be synthesized by ultrasonic reduction of Ag precursors. This ultrasonic preparation method can improve the yield of nanoparticles with strong repeatability.

1.2.5 Microwave Method

In the past 25 years, microwave chemistry has gradually became a mature technology from lab-to-fab. Microwave heating is a polarization phenomenon with a frequency range of 0.3–300 GHz. In the electromagnetic spectrum, the microwave frequency is between infrared and radio. Microwave chemistry mainly depends on two mechanisms—the dipole mechanism and conductance mechanism. The condition for the dipole mechanism is that polar molecules will become aligned with the electric field orientation under a high-frequency electric field. When it happens, molecules release enough heat to drive the reaction. In the conductance mechanism, the irradiated sample is a conductor. The current carrier (electrons, ions, etc.) passes through the material and leads to polarization under the electric field. Meanwhile, the accompanying polarization current will heat the sample with the assistance of electric field.

In 1986, microwave chemistry was applied to the chemical industry, especially in the preparation of nanomaterials [62–65]. The rapid consumption of feeds in microwave chemical reactions reduces the aggregation of nanoparticles, thus nanocrystals hold a narrow size distribution. Zaleski et al. prepared Ag nanowires utilizing NaCl and $AgNO_3$ as precursors in ethylene glycol and PVP solution by microwave method [66]. The preparation of Ag nanowires is highly dependent on the microwave heating energy, time and the ratio of NaCl and $AgNO_3$. The authors found that the increasement of heating time would lead to nanowire fusion or shorter Ag nanowires. Bhalla et al. developed a similar strategy to synthesize Ag and Pt nanocrystals with ethylene glycol, PVP and NaOH by microwave method [67]. In this method, nanospheres rather than nanowires are obtained by using NaOH. Komarneni et al. also prepared self-assembled hexagonal Ag nanoparticles in a one-pot microwave-assisted solvothermal method [68].

In addition, it is possible to large-scale prepare size-controlled Ag nanoparticles in $AgNO_3$ and sodium citrate aqueous solutions using formamide as a reductant under microwave irradiation [69]. Besides Ag and Pt metal nanocrystals, microwave method can also be utilized to prepare Au, Pd, Ru and Rh alloys with controllable morphology and size [70–73] (Fig. 1.7). Therefore, microwave irradiation has rapidly became a widely accepted method of chemical synthesis.

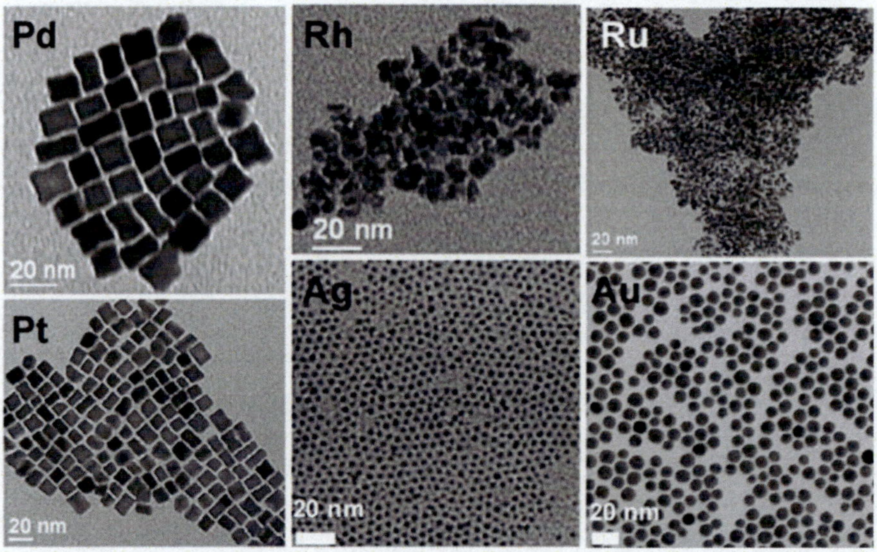

Fig. 1.7 Noble metal nanoparticles prepared by microwave method [70]

1.3 Characterizations of Noble-Metal Nanocrystal

Comprehensive characterization of noble-metal nanocrystal is an important indicator for rationally synthesizing morphology-controlled nanocrystals. It is an essential method to demonstrate the successful preparation and explain the evolution of nanoparticle morphology and optimize synthesis schemes. Characterizations mainly include components, morphology, dimensions, physical (e.g., optical and spectral properties) and chemical properties (e.g., surface charge and functionality). The size (e.g., diameter, length, width and height), volume, aspect ratio, and surface roughness of the material are also typical parameters required to be considered. At present, there exist many routines that can be used to characterize nanoparticles. Typical characterization techniques include X-ray diffraction (XRD), X-ray photoelectron spectroscopy (XPS), atomic force microscope (AFM), scanning tunnel microscope (SEM) and transmission electron microscope (TEM). The following part briefly describes several representative characterization techniques.

TEM is a powerful tool to obtain morphology and microstructures of nanocrystals [74, 75]. Low-magnification TEM can be employed to observe the size and purity of nanocrystals. However, TEM can only provide part of the information about the morphology of nanocrystalline. For example, the truncated octahedron with exposed {111} and {100} crystal surfaces show a square outline from the [100] crystal axis. However, the square outline may be a cube or a square diamond cone [76]. Therefore, other techniques should be combined to further confirm the real morphology of nanocrystalline. High-resolution TEM electron diffraction and region-selecting electron diffraction can provide atomic structure information (lattice spacing, lattice exposure, etc.), providing more details about the atomic structure of nanocrystalline. Atomic Force Microscope (AFM) can characterize the thickness, morphology and surface force of nanoplates [77]. AFM technology of contact mode (scanning probes are in contact with nanocrystalline surfaces) can obtain information about the lattice resolution, while AFM technology off contact mode (scanning probes do not come into contact with nanocrystalline surfaces) can only obtain information of atomic resolution. Unlike other microscope techniques, AFM can characterize the individual self-assembling and nanoparticle clusters of dry or wet at room temperature. Scanning electron microscope (SEM), field emission scanning electron microscope (FE-SEM) and ultra-high-resolution FE-SEM can characterize the surface of samples and obtain high-resolution scanning photos. X-ray diffraction (XRD) can infer the diameter of particles based on the width and narrowness of the diffraction peak, as well as confirm the sample crystal phase based on the peak position of the diffraction peak and the characteristic peak [78–80]. X-ray photoelectronic spectrum (XPS) is a beneficial tool for detecting the electron structure and surface composition of nanocrystalline surfaces [81–83]. This technique utilizes the photoelectronic energy excited from the inner electrons of the catalyst surface or near-surface atoms (Fig. 1.8).

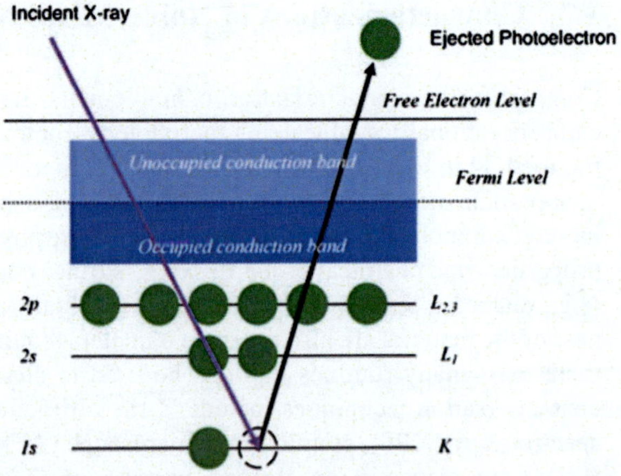

Fig. 1.8 Diagram of the photoelectronic emission process [81]

1.4 The Application of Noble Metal Nanocrystalline

Nanocrystalline catalysis shows a rapid growth, attracting extensive research interest in the industrial catalysis, chemical synthesis and biopharmaceutical industries. The application of noble metal nanocrystals in organic catalysis, electrocatalysis and surface-enhanced Raman (SERS) are described below.

1.4.1 The Application of Noble Metal Nanocrystals in Organic Catalysis

The major difference between heterogeneous and homogeneous catalysis is that reactant molecules should first be adsorbed to the surface of the nanocrystalline catalyst. Since the surface-volume ratio of nanocrystalline height is very attractive for catalytic applications, nanocrystalline catalysis is essentially a surface-exposed and atom-involved reaction, that is, surface interface catalysis. Therefore, researchers have studied the relationship between the surface structure and catalysis of nanocrystalline by regulating the atomic arrangement of the nanocrystal interface. For example, our group reported a common method to control the Pt-Ni alloys with different facets exposure. The relationship between nanocrystal structure and catalytic activity was explored by the model reaction of styrene and nitrobenzene hydrogenation [13] (Fig. 1.9).

In addition, our group also explored the selective hydrogenation performance of Ru with different crystal phases. It was found that fcc-Ru and hcp-Ru differed greatly in nitrobenzene hydrogenation and styrene hydrogenation, which may be due to the different atoms arrangement of different crystal phases [84] (Fig. 1.10).

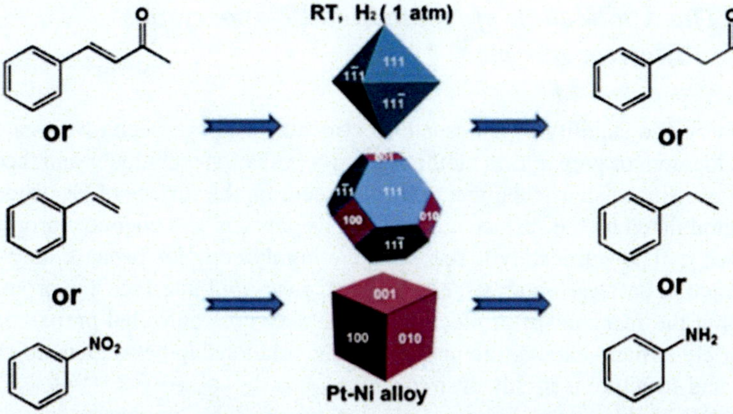

Fig. 1.9 Schematics of Styrene and nitrobenzene hydrogenation reaction using Pt-Ni alloy catalysts [13]

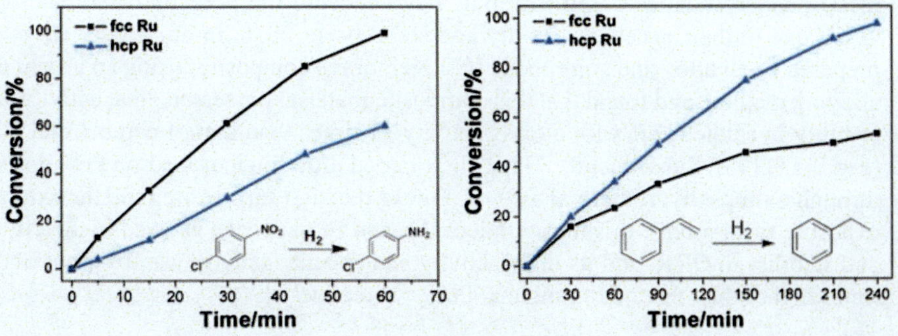

Fig. 1.10 Hydrogenation reactions of fcc-Ru and hcp-Ru [84]

Moreover, noble metal nanocrystals can also be used for other types of organic catalysis, such as selective oxidation [85], coupling reactions [86], C-H activation [87], amine oxidation [88], tandem reactions [89], and so on. For example, Narayanan et al. found that Pd nanocrystalline could catalyze iodized benzene and benzene boric acid to obtain biphenyls, and the catalytic activity was closely related to the surface exposed facets of Pd nanocrystalline [90]. Shiraishi et al. found that Ag sols could oxidize ethylene selectively into ethylene oxide in ethanol/water solvents with high activity and selective [91]. El-Sayed et al. studied the activity of Pt nanocrystals in electron transfer reactions. They studied the Pt nanocrystals with tetrahedron, cubes, and spherical morphology in the electron transfer reactions of $[Fe(CN)_6]^{3-}$ and $S_2O_3^{2-}$ to form $[Fe(CN)_6]^{4-}$ and $S_4O_6^{2-}$. It was found that the average rate constant increased as the percentage of corners and edges atoms of the Pt nanocrystalline surface increased [92].

1.4.2 *The Application of Noble Metal Nanocrystals in Electrocatalysis*

The activity and stability of nanoscale electrocatalysts have been investigated with the development of preparation methods, characterization techniques and theoretical calculation. Since electrochemical reactions occur on the surface of electrodes, they can be modulated by the surface area as well as physical and chemical properties of electrodes [93]. Electrocatalytic reactions are highly complex because they involve the interaction between catalysts, electrolytes, gas/liquid interface and products. To understand the mechanism of electrocatalytic activity, controlled preparation of a wide variety of nano-catalysts to understand the relationship between nano-catalytic activity and structure is highly desired [94].

Currently, noble metal nanocrystals can be used for oxygen reduction reaction (ORR), carbon dioxide reduction (CO_2RR), ethanol/methanol electrooxidation, oxygen extraction reaction (OER), hydrogen evolution reaction (HER), ammonia synthesis, etc. Among the widely applied in energy-conversion reactions of noble metal nanocrystals, Pt-based materials have become the potential candidates for ORR, due to their excellent activity and stability. In 2016, in our group, He et al. prepared Pt-Ni alloy and amorphous NiB membrane composites using an in-situ de-alloying method, and found that this composite material possessed great activity and stability in acidic ORR with a mass activity 27 times than that of commercial Pt/C (Fig. 1.11) [95]. Subsequently, He et al. prepared ultrathin icosahedron Pt nanocage through a simple hydrothermal method. Due to the high surface area and the extreme exposure of Pt atoms, this ultrathin icosahedron Pt nanocage showed high activity and stability in ORR, and its mass activity and specific activity was 10 times and 7 times higher than that of commercial Pt/C, respectively [96].

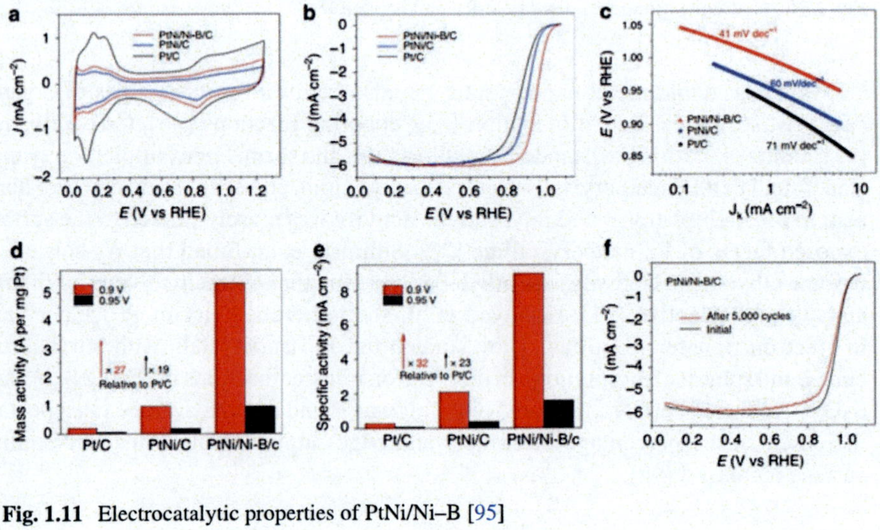

Fig. 1.11 Electrocatalytic properties of PtNi/Ni–B [95]

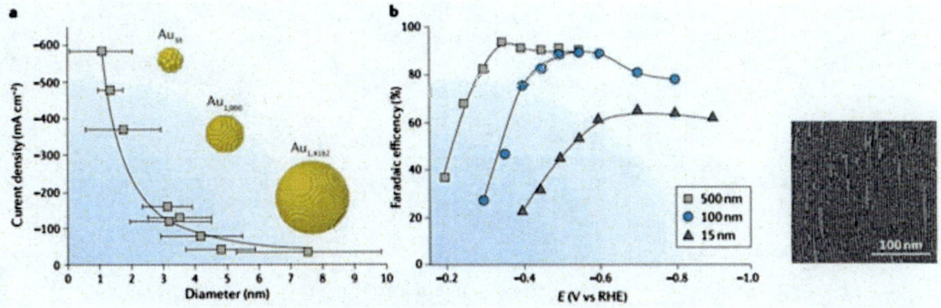

Fig. 1.12 Au nanoparticles used for CO_2RR [99]

The Commonly used metal nanomaterials for CO_2RR are Au, Pd, Cu and Sn. Zhu et al. studied Au nanoparticles (4–10 nm) and nanowires (15–500 nm) for CO_2RR [97, 98]. They found that larger Au nanoparticles or longer Au nanowires could selectively convert CO_2 to CO. Other studies also found that when the size of Au was less than 5 nm, the CO_2RR activity is higher due to the increase of unsaturated sites on the nanoparticles surface [99] (Fig. 1.12).

In addition, noble metal nanocatalysts are also widely used in catalyzing OER and the commonly used OER catalysts are mainly Ru/Ir-based materials. For example, Tilley et al. synthesized Pd/Ru nanocrystals with a nanosized branch structure and low-index facets of Ru through controlled-growth. Strikingly, this special structure imparted Pd/Ru with good activity and stability in acidic OER, and only 225 mV overpotential was required to drive the current density of 10 mA/cm^2 [100] (Fig. 1.13).

HER, as one of the most important electrocatalytic reactions, is the cathode reaction of water splitting, which could produce clean energy of H_2. Pt, Pd and Ru are extensively utilized in HER. In 2017, our group prepared atomically dispersed copper–platinum dual sites alloyed with palladium nanorings (Pd/Cu-Pt NRs)

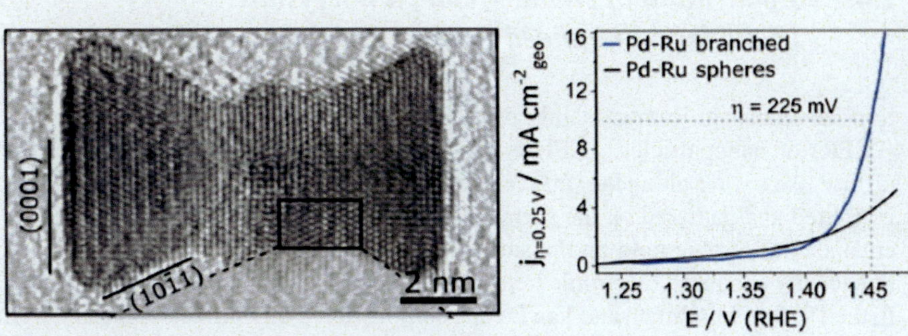

Fig. 1.13 OER activity of Pd/Ru catalysts [100]

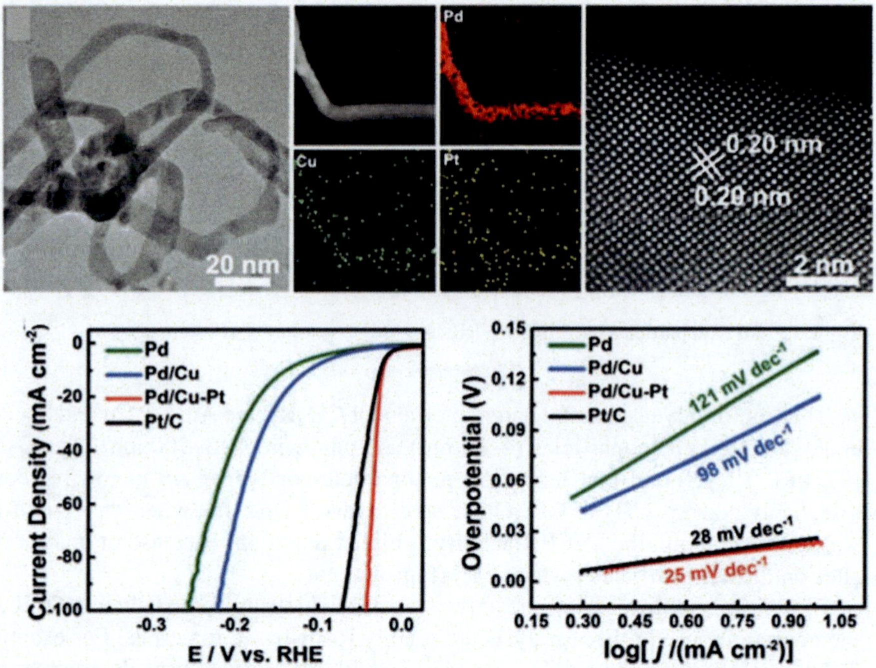

Fig. 1.14 HER activity of Pd/Cu-Pt nanorings [101]

through a two-step approach [101]. The Pd/Cu-Pt NRs exhibited great HER properties in 0.5 M H_2SO_4 with an overpotential of 22.8 mV at a current density of 10 mA/cm^2, exceeding that of the commercial Pt/C (30 mV) (Fig. 1.14).

1.4.3 Application of Noble Metal Nanocrystals in Surface-Enhanced Raman Scattering (SERS)

Surface-enhanced Raman mainly depends on localized surface plasmon resonance (LSPR) of nanoparticles. LSPR refers to the technique where the charge cannot diffuse like a wave along the surface due to the inducement of nanoparticles, and thus is limited and gathered on the surface. Accordingly, LSPR could result in a locally enhanced electronic field on the surface of nanoparticles. The direct application of this locally enhanced electronic field is surface-enhanced Raman scattering (SERS). In the 1970s, Fleischman and Van Dyune demonstrated this surface-enhanced Raman scattering technology for the first time.

Regulating the plasma properties of nanocrystals such as Ag and Au can enhance the sensitivity of SERS, which is a useful technique for molecular sensing. Each organic molecule has its own unique Raman spectrum, and the enhancement of

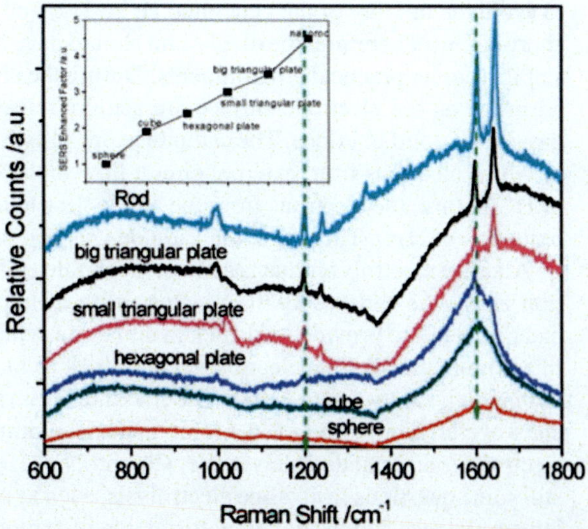

Fig. 1.15 SERS spectra of Ag nanoparticles with different morphology [102]

signals is necessary for the detection of low-concentration target. The strong local electronic field generated by LSPR can enhance the Raman signal of molecules adsorbed to metal nanocrystals by several orders of magnitude. It was found that the position and strength of LSPR peaks can be controlled by regulating the morphology of metal nanocrystals. For example, Yang et al., synthesized Ag nanoparticles with different shapes (e.g., triangle, hexagon, nanorod, nanoplates, cube, and polyhedron) by a solvothermal method. These Ag nanoparticles with different shapes exhibited discrepant surface-enhanced Raman scattering [102], roughly following the order of: nanorod > triangular nanoplate > hexagonal nanoplate > cube > sphere (Fig. 1.15).

1.5 Electrocatalytic Water Splitting Technology

1.5.1 Introduction of Electrocatalytic Water Splitting Technology

Water splitting could convert water molecules into hydrogen (H_2) and oxygen (O_2) through sustainable electricity, that is a process of converting electricity into chemical energy. During this process, current flows between the two electrodes that are separated and immersed in the electrolyte. To meet the industrial demand, electrodes require to possess several characteristics, such as corrosion resistance, great conductivity, high activity and durability. Meanwhile, the electrolyte requires to keep invariable during the reaction, and also could not react with the electrode. Electrode, diaphragm and electrolyte are crucial components in electrolyzer. Among the significant components that the diaphragm plays a key role in separating the two electrodes to avoid the mixture of the generated H_2 and O_2 and also prevents the occurrence of short-circuit. Therefore, the diaphragm should have high ionic conductivity and high stability under practical environments. During the electrolysis process, the reactants adsorbed on the electrode surface are acquire/release electrons, thereby forming a gas–liquid-solid interface. The cathode happens hydrogen reduction reaction (HER), of which electrons from external circuit flow to the cathode to induce the reduction of H^+. In turn, the electrons from the anode flow to the external circuit to induce the oxidation of H_2O. Therefore, the cathode gets H_2, while the anode generates O_2.

Alkaline electrolyzer that is a mature technology, is widely utilized for the production of H_2. As early as 1920, a factory with a H_2-producing capacity of 100 MW had been built to provide feedstock in ammonia synthesis and petroleum processing. In addition, the alkaline electrolyzer can work as long as 15 years, and also operate with a high safety factor [103]. The maximum current density of a typical alkaline electrolyzer can be up to 0.4 A/cm^2 under operating conditions, i.e., 80 °C, KOH electrolyte, and anode catalyst (Fe, Co and Ni) [104–106]. Nevertheless, there are still some questions in alkaline electrolysis, such as small current densities due to the large resistance and the potential formation of explosive gas mixtures in the reaction [107]. Acidic electrolyzer that depends on proton exchange membrane (PEMWEs), is just developed in recent years. Its feature is to use the solid polymer membrane (thickness < 0.2 mm) that only allow H^+ to pass while preventing OH^- and gas from going through. The commercial electrodes for PEMWE often employ Pt as the cathode and Ir/IrO_2 as the anode with a noble metal loading in the range of 0.05–2 g/cm^2.

1.5.2 Oxygen Evolution Reaction (OER)

1.5.2.1 The Introduction of OER

An electron-donating counter reaction is highly desired in fuel molecular conversion and storage processes, such as the production of H_2 from water electrolysis and hydrocarbons from CO_2RR. Electrocatalytic water oxidation to oxygen, that is oxygen evolution reaction (OER), is emerging as one of the most potential reaction to realize the above goal [108]. OER is not only a key step in electricity storage, but also plays a pivotal role in other electrochemistry processes, such as electrochemistry deposition [109]. As known, OER is a complex reaction associated with multi-step electrons transfer, even though the best commercial catalyst still requires a high overpotential to drive the reaction in the practical condition, consequently reducing the efficiency. In addition, the harsh and oxidative OER environment would lead to the poor stability of the catalyst. According to the electrolyte pH, OER is divided into acidic OER and alkaline OER. Compared with acidic OER [110], the requirements of alkaline OER catalysts are relative ordinary due to its mildly operating conditions, and the development of designing catalyst is mature [111, 112]. Therefore, the following section will focus on the introduction of acidic OER catalysts as well as the challenges faced by acidic OER.

1.5.2.2 Oxygen Evolution Reaction Catalysts

(1) Single metal Ru/Ir oxide

The electrochemical OER activity trend of metal oxides is Ru > Ir > Rh > Pt > Au [113–116], which is evaluated by the overpotential to reach a current density of 5 mA/cm^2. Osmium (Os) is found to be the most active OER catalyst, while Os shows extreme poor stability in industrial electrolyzer under acidic OER conditions, and thus Os cannot be referred to a "catalyst". Cherevko et al. found that the stability trend is Pt > Rh > Ir > Au ≥ Ru [115]. By comparing the activity and stability, it can be conclude that the activity and stability of catalysts show the opposite trend, and OER catalysts with high activity are accompanied by poor stability.

Various approach can be used for the preparation of Ru/Ir oxides, such as electrochemical preparation [117, 118], thermal treatment of metal precursors [119], physical vapor deposition and decomposition of suitable precursors [120]. Apart from the metal cations existing in the oxide, the synthesis conditions also determine the OER performance. Generally, the oxides prepared by thermodynamical method exhibit higher stability, yet lower activity than the electrochemically prepared oxides. For example, the stability of Ru and Ir oxides prepared by thermodynamical method are 2–3 orders of magnitude higher than that of electrochemically prepared oxides [119].

(2) Ru-Ir mixed oxide

To concurrently reduce the amount of noble metal and improve the activity and stability, researchers are often prepared Ru-Ir mixed oxide to replace the Ru/Ir oxide. Exner et al. suggested that partial replacement of the cation in an oxide could regulate the adsorption energy of the intermediate and thus increase the activity [121]. Many studies have found that Ru-Ir mixed oxide showed lower activity, but higher stability than Ru oxide. However, Ru-Ir mixed oxide is more active than Ir oxide, while lower stability than Ir oxide [122, 123]. Accordingly, the OER activity and stability can be well adjusted by regulating the mixing ratio between those of single metal oxides.

(3) Metal doping

Although the OER activity and stability of Ru-Ir mixture can be improved through optimizing the chemically mixing of single metal oxides from a functional perspective. Both Ru and Ir belong to the Pt-group metals and are rare, which necessitate to develop other cheap metals doping to low the amount of Ru/Ir. Recent studies reported that the OER performance of Ru and Ir oxides can be optimized by Ni [124–126]. Macounová et al. developed a spinel Ru-Ni mixed oxide containing Ni clusters and abundant defects (Ni content as high as 30%), with a higher activity than that of Ru oxide [127]. To reduce the use of Ir, Nong et al. prepared IrO_x@IrNi

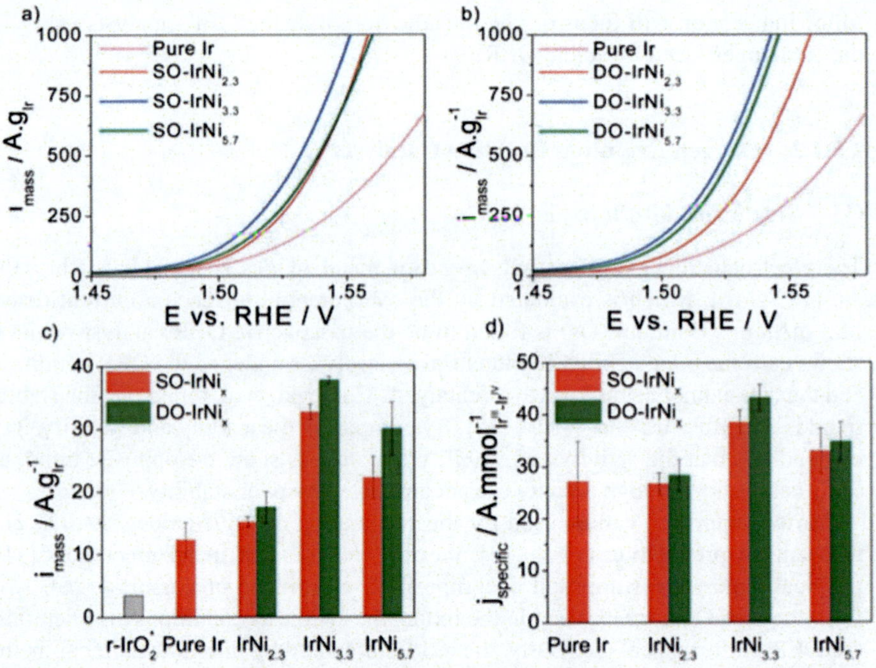

Fig. 1.16 Acidic OER performance of Ir-Ni nanocrystals with different amounts of Ni [128]

core–shell nanoparticles [128], and when the Ni content was approximately 77 at%, the OER activity was the optimium (Fig. 1.16).

Except Ni doping, other metal cations are also employed to optimize the OER performance of Ru and Ir oxides. For Ru oxides, their OER performance can be improved by mixing with La [129], Ce [130], Ti [131], Nb [132], Pb [133], Sn [134] and Pt [135]. For Ir oxides, their OER performance can be boosted by mixing with Sn [136], Sb [137] and Co [138]. For example, Petrykin et al. reported a Ru-Zn mixed oxide and found that the addition of Zn greatly improved the selectivity of OER [139]. Petrykin et al. also found that Ru-Co mixed oxide had a similar improvement in the selectivity of OER [140].

In addition, OER performance can be finely regulated by changing the size and morphology of catalysts. By adjusting the morphology of catalysts, the exposed facets can be changed, thereby modulating the adsorption energy of the intermediate, and eventually optimizing the catalytic performance. Lim et al. found that dendritic Ir, Ir-Ni and Cu-Ir nanocages with controllable morphology exhibited better OER performance than spherical Ir [141] (Fig. 1.17).

(4) Support

Support-like material is a promising candidate for OER due to its high specific surface area, excellent conductivity and strong corrosion resistance. For example,

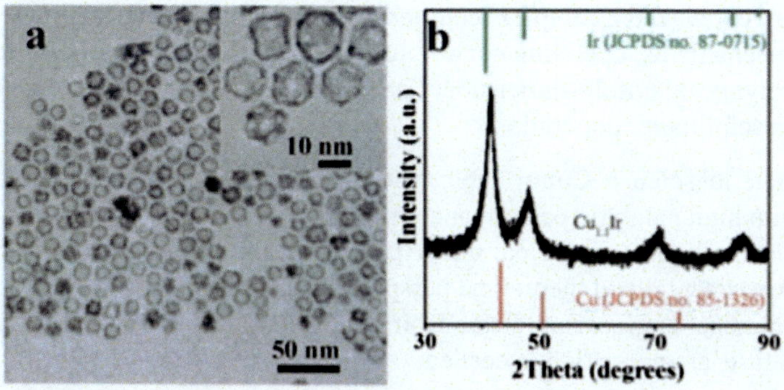

Fig. 1.17 Characterizations of $Cu_{1.11}Ir$ nanocage [141]

carbon-based materials (e.g., carbon black, nanotubes, nanofibers), porous carbon and boron-doped diamond are all commonly used supports.

In addition, Sn-based, In-based, W-based and Ti-based supports with high conductivity are often considered, i.e., Sb-doped SnO_2 (ATO), F-doped SnO_2 (FTO) and Sn-doped In oxide (ITO). Strasser et al. demonstrated ATO that can show good conductivity and high stability, is an excellent support for dendritic Ir nanocrystals and IrO_x@IrNi core–shell nanocrystal. Also, ATO and ITO as the supports of OER catalysts, have been proven to applicable in industrial electrolyzer due to their excellent stability [142].

1.5.2.3 The Development and Challenge of OER

OER catalysts largely hamper the large-scale application of electrochemical energy conversion and storage technology, which increase the cost of the device. The disadvantages of OER catalysts are as follows:

(1) Activity. For OER reaction, even though the most active catalyst still needs a high overpotential to drive the reaction, leading to excessive energy consumption.
(2) Stability. Since the OER environment is particularly harsh, this place great emphasis on the requirement of high stability for the catalyst. At present, the stability of OER catalysts is not able to satisfy the long-term operation of the electrolyzer.
(3) Cost and reserves. Ru or Ir-based materials are commonly applied in OER, which limits the large-scale promotion of electrolyzer, owing to their high cost and low reserves. Therefore, it is imperative to search a cheap catalyst with abundant reserves to replace Ru/Ir for OER.

The advancement in the development of catalysts can satisfy the industrial demand to some extent, however, it is still challenging to design a catalyst with the above three advantages at the same time. To promote the development of this discipline, it is necessary to reveal the complex relationships between the material properties, degradation mechanisms, operating conditions, and the real structure of electrodes. Obviously, solving this multi-dimensional problem inevitably requires the combination of interdisciplinary cooperation.

(1) Material science. Controllable synthesis and characterization of nano-catalysts with high catalytic performance is needed.
(2) Electrochemistry. A clear understanding of the electrochemical process from experimental and theoretical perspectives is needed.
(3) Chemical engineering. Establishing the relationship between lab research and industrial application is needed.

The perfect combination of materials science, electrochemistry and chemical engineering can promote the development of OER at an industrial-scale.

1.5.3 Electrocatalytic Hydrogen Evolution Reaction (HER)

1.5.3.1 Introduction of Electrocatalytic Hydrogen Evolution Reaction

Electrocatalytic water splitting involves two half reactions: oxygen evolution reaction (OER) and hydrogen evolution reaction (HER). As the cathodic reaction of water splitting, HER can be classified into acidic HER and alkaline HER according to the electrolyte. In the following section, we mainly discuss about the acidic HER due to its potential industrial application.

In acidic media, HER generally involves three possible steps [143] (Fig. 1.18). First, it was triggered by the Volmer process ($H^+ + e^- \rightarrow H_{ads}$), during which H_{ads} is generated and adsorbed on the electrode surface. Subsequently, the adsorbed H atom follows the Tafel process ($2H_{ads} \rightarrow H_2$) and/or the Heyrovsky process ($H_{ads} + H^+ + e^- \rightarrow H_2$) to produce H_2. Given the necessity of Volmer process in HER, the free energy of hydrogen adsorption (ΔG_H) becomes a vital descriptor in evaluating HER activity [144]. For instance, the ΔG_H of Pt is close to 0, suggesting that Pt is one of the best-performing HER catalysts. More positive ΔG_H represents stronger H adsorption on electrode surface, leading to easier Volmer process, while more negative ΔG_H results in weaker H_{ads} adsorption and harder Volmer process. As a result, a promising HER electrocatalyst highly requires a suitable ΔG_H.

Besides, the HER activities of electrocatalysts are also determined by the intrinsic activity, Tafel slope, stability, Faraday efficiency as well as TOF.

(1) Intrinsic activity. The intrinsic activity of electrocatalysts can be evaluated by CV and LSV measurements. To investigate the electrocatalytic activity, we can examine the steady-state current recorded at different time intervals (intervals: 5 min) under several potentials, which is further normalized to the electrode surface area or catalyst mass. The intrinsic activity of catalysts can be compared by the overpotential to drive a current density of 10 mA/cm^2.

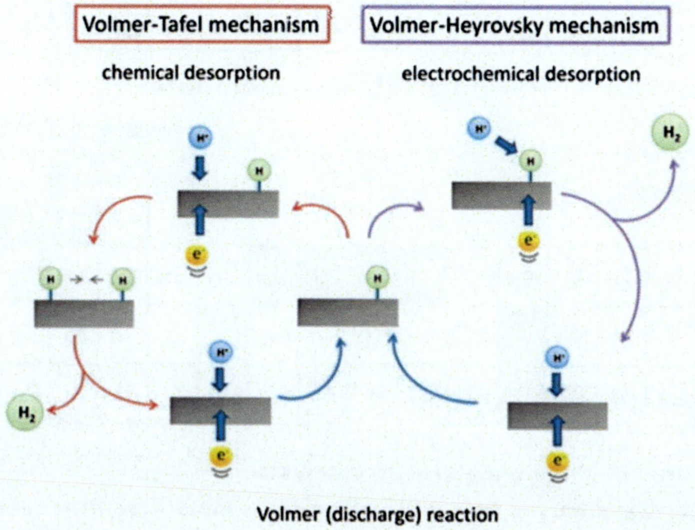

Fig. 1.18 The mechanism of HER in acidic medium [145]

(2) Tafel slope. Tafel slope describes the relationship between steady state current density and overpotential. The Tafel slope derives from the formula: $\eta = a + b \cdot \log^{j}$, in which b represents the Tafel slope. Generally, efficient HER catalyst requires high j_0 and low Tafel slope (b).

(3) Stability. Since HER operates under strong reductive environment, the HER catalyst with great stability is crucial for industrial application. The catalytic stability can be evaluated in two ways: (1) tracing the current variation with time (i–t curve), of which the current density over 10 mA/cm^2 and the time over 10 h are required. (2) measuring CV or LSV for several cycles (over 5000 cycles).

(4) Faraday efficiency. The Faraday efficiency in HER refers to the ratio of H_2 generated and theoretical value.

(5) Turnover frequency (TOF). TOF is defined as the intrinsic activity of per active site.

1.5.3.2 Electrocatalysts for Hydrogen Evolution Reaction

HER catalysts can be classified into three groups according to their physical and chemical properties: (i) noble metal Pt—the best-performing HER catalyst; (ii) non-noble metal electrocatalysts, include Fe, Cu, Co, Mo, Ni and W; (iii) non-metallic material, i.e., B, P, C, S, N, Se.

By comparing the abundance of the above-discussed catalysts, it can be concluded that: (i) The abundance of Pt is around $3.7 \times 10^{-6}\%$, which is lower than that of other non-noble metals for several orders of magnitude. As a result, Pt is one of the most

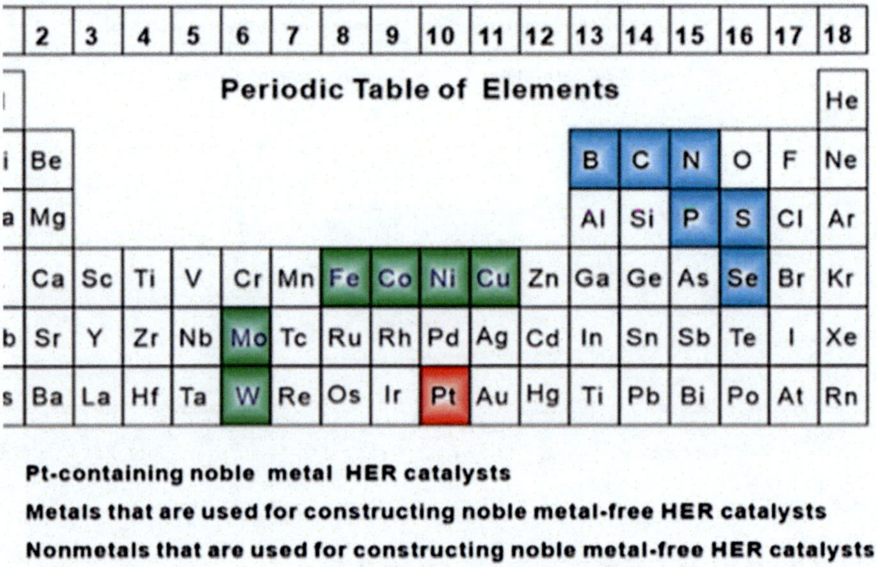

Fig. 1.19 HER electrocatalysts [143].

precious metal for HER; (ii) The abundance order of non-noble metals is W = Mo < Co < Cu < Ni < Fe (Fig. 1.19).

1.5.3.3 The Development and Challenge for Electrocatalytic Hydrogen Evolution Reaction

Noble metal Pt is the best-performing HER catalyst, however, its high cost and limited reserve hinder the large-scale application. Although some progress has gained in simultaneously improving Pt mass activity and reducing Pt dosage, it is still challenging to realize the industrialization of water splitting.

(1) Mechanism exploration: The exploration of mechanism is of significance to optimizing and designing HER catalysts. To date, HER mechanism at atomic level, especially for complex compound catalysts still remain ambiguous, which urgently requires theoretical calculation combining with in-situ characterization to clarify the underlying catalytic process.
(2) Standard measurement: Establishing standard measurement methods are of great importance to evaluate and screen HER catalysts prepared at different conditions. At present, the discrepancies in catalyst dosage, preparation process and reaction condition (e.g., electrolyte) result in directly comparing the catalytic activity difficult. For instance, the current measured in HER is often normalized to electrode area without considering the catalyst dosage. Actually, the current is closely related with the catalysts loading, thus the current

normalized to electrode area may lead to inadequate assessment in some cases. To gain a better understanding of catalysts performance, researchers need to provide more details, including Tafel slope, TOF, catalyst dosage, Faraday efficiency, stability. Based on the above discussion, researcher should focus on as follows.

i. Searching new HER catalysts. Exploring new HER materials is one of our central goals. We intended to prepare an ideal non-Pt electrocatalyst with the following advantages: (a) comparable HER activity to Pt; (b) high durability as long as several years; (c) high chemical/catalytic stability in wide pH ranges; (d) Low cost; (e) easy large-scale production. To date, HER catalyst with all the above advantages has not been reported.
ii. the HER electrocatalyst compatible with OER electrocatalyst. The efficiency of water splitting is not only determined by HER and OER, but also closely related to the compatibility of two half reactions. Therefore, the evaluation of HER catalysts should be conducted in practical water splitting.

1.6 The Research Content and Purpose

Noble metal nanocrystal is widely used in chemical, energy, medical and other fields, due to their special physical and chemical properties. Although some advancement has been achieved in regulating the size and morphology of nanocrystal, the controllable preparation of ternary nanocrystals still remains a great challenge. Besides, since precious metal catalysts play a vital role in energy storage and conversion, the biggest challenge at present is to develop advanced electrocatalysts and thereby realize the widespread promotion of clean energy technologies. Therefore, our research is focused on the controllable preparation of ternary nanocrystals and the exploration of their applications:

We report an epitaxial-growth-mediated method to grow face-centered cubic (fcc) Ru, which is thermodynamically unfavorable in the bulk form, on the surface of Pd–Cu alloy. Induced by the galvanic replacement between Ru and Pd–Cu alloy, a shape transformation from a Pd–Cu@Ru core–shell to a yolk–shell structure was observed during the epitaxial growth. The successful coating of the unconventional crystallographic structure is critically dependent on the moderate lattice mismatch between the fcc Ru overlayer and $PdCu_3$ alloy substrate. Further, both fcc and hexagonal close packed (hcp) Ru can be selectively grown through varying the lattice spacing of the Pd–Cu substrate. The presented findings provide a new synthetic pathway to control the crystallographic structure of metal nanomaterials.

We have synthesized a series of $PtCu_x/Pt_{skin}$ core–shell structures with atomically dispersed Ru_1 and have unravelled their mechanism of formation, oxidation resistance and the origin of the enhanced OER catalysis by exhaustively examining their structures, coordination environments and oxidation states. Through sequential acid etching and electrochemical leaching, the structure of $PtCu_x$ alloys can be varied (to give $PtCu_3$, PtCu and Pt_3Cu), which modulates effectively the OER activity catalysed by the Ru_1. The best catalyst, Ru_1–Pt_3Cu, delivers 220 mV overpotential to achieve a current density of 10 mA cm^{-2} for acidic OER, with ten times longer lifetime over commercial RuO_2. We found that there is a volcano-type relation between the OER activity and the lattice constant. We argue that the compressive strain of the Pt_{skin} shell effectively engineers the electronic structure and redox behaviour of single atomic Ru anchored at the corner or step sites of the Pt-rich shell, with optimized binding of oxygen intermediates and better resistance to over-oxidation and dissolution.

Submonolayer Pt was controllably deposited on an intermetallic Pd_3Pb nanoplate (AL–Pt/Pd_3Pb). The atomic efficiency and electronic structure of the active surface Pt layer were largely optimized, greatly enhancing the acidic HER. AL–Pt/Pd_3Pb exhibits an outstanding HER activity with an overpotential of only 13.8 mV at 10 mA/cm^2 and a high mass activity of 7834 A/g_{Pd+Pt} at − 0.05 V, both largely surpassing those of commercial Pt/C (30 mV, 1486 A/g_{Pt}). In addition, AL–Pt/Pd_3Pb shows excellent stability and robustness. Theoretical calculations show that the improved activity is mainly derived from the charge transfer from Pd_3Pb to Pt, resulting in a strong electrostatic interaction that can stabilize the transition state and lower the barrier.

References

1. Fahlman BD (2007) Semiconducting materials. In: Materials chemistry. Springer, Dordrecht, pp 153–219
2. Zijlstra P, Chon JW, Gu M (2009) Five-dimensional optical recording mediated by surface plasmons in gold nanorods. Nature 459:410–413
3. Elghanian R, Storhoff JJ, Mucic RC, Letsinger RL, Mirkin CA (1997) Selective colorimetric detection of polynucleotides based on the distance-dependent optical properties of gold nanoparticles. Science 277:1078–1081
4. Chen C-D, Cheng S-F, Chau L-K, Wang CC (2007) Sensing capability of the localized surface plasmon resonance of gold nanorods. Biosens Bioelectron 22:926–932
5. Anker JN, Hall WP, Lyandres O, Shah NC, Zhao J, Van Duyne RP (2010) Biosensing with plasmonic nanosensors. Nanosci Technol: Collect Rev Nat J 308–319
6. Sales EA, Benhamida B, Caizergues V, Lagier J-P, Fiévet F, Bozon-Verduraz F (1998) Alumina-supported Pd, Ag and Pd–Ag catalysts: preparation through the polyol process, characterization and reactivity in hexa-1, 5-diene hydrogenation. Appl Catal A: General 172:273–283
7. Astruc D, Lu F, Aranzaes JR (2005) Nanoparticles as recyclable catalysts: the frontier between homogeneous and heterogeneous catalysis. Angew Chem Int Ed 44:7852–7872
8. Matsui I (2005) Nanoparticles for electronic device applications: a brief review. J Chem Eng Jpn 38:535–546
9. Crespo-Biel O, Ravoo BJ, Huskens J, Reinhoudt DN (2006) Writing with molecules on molecular printboards. Dalton Trans 2737–2741
10. Stamenkovic VR, Fowler B, Mun BS, Wang G, Ross PN, Lucas CA, Marković NM (2007) Improved oxygen reduction activity on Pt3Ni (111) via increased surface site availability. Science 315:493–497
11. Modeshia DR, Walton RI (2010) Solvothermal synthesis of perovskites and pyrochlores: crystallisation of functional oxides under mild conditions. Chem Soc Rev 39:4303–4325

12. Cushing BL, Kolesnichenko VL, O'Connor CJ (2004) Recent advances in the liquid-phase syntheses of inorganic nanoparticles. Chem Rev 104:3893–3946
13. Wu Y, Cai S, Wang D, He W, Li Y (2012) Syntheses of water-soluble octahedral, truncated octahedral, and cubic Pt–Ni nanocrystals and their structure–activity study in model hydrogenation reactions. J Am Chem Soc 134:8975–8981
14. Yang Y, Matsubara S, Xiong L, Hayakawa T, Nogami M (2007) Solvothermal synthesis of multiple shapes of silver nanoparticles and their SERS properties. J Phys Chem C 111:9095–9104
15. Tan Y, Dai X, Li Y, Zhu D (2003) Preparation of gold, platinum, palladium and silver nanoparticles by the reduction of their salts with a weak reductant–potassium bitartrate. J Mater Chem 13:1069–1075
16. He H-X, Zhang H, Li QG, Zhu T, Li S, Liu Z-F (2000) Fabrication of designed architectures of Au nanoparticles on solid substrate with printed self-assembled monolayers as templates. Langmuir 16:3846–3851
17. Murphy CJ, Sau TK, Gole AM, Orendorff CJ, Gao J, Gou L, Hunyadi SE, Li T (2005) Anisotropic metal nanoparticles: synthesis, assembly, and optical applications. J Phys Chem B 109:13857–13870
18. Sau TK, Murphy CJ (2004) Room temperature, high-yield synthesis of multiple shapes of gold nanoparticles in aqueous solution. J Am Chem Soc 126:8648–8649
19. Habas SE, Lee H, Radmilovic V, Somorjai G, Yang AP (2007) Shaping binary metal nanocrystals through epitaxial seeded growth. Nat Mater 6:692–697
20. Millstone JE, Métraux GS, Mirkin CA (2006) Controlling the edge length of gold nanoprisms via a seed-mediated approach. Adv Func Mater 16:1209–1214
21. Liu X, Huang R, Zhu J (2007) Functional faceted silver nano-hexapods: synthesis, structure characterizations, and optical properties. Chem Mater 20:192–197
22. Yener DO, Sindel J, Randall CA, Adair JH (2002) Synthesis of nanosized silver platelets in octylamine-water bilayer systems. Langmuir 18:8692–8699
23. Wei G, Zhou H, Liu Z, Song Y, Wang L, Sun L, Li Z (2005) One-step synthesis of silver nanoparticles, nanorods, and nanowires on the surface of DNA network. J Phys Chem B 109:8738–8743
24. Zhang J, Liu H, Wang Z, Ming N (2007) Shape-selective synthesis of gold nanoparticles with controlled sizes, shapes, and plasmon resonances. Adv Func Mater 17:3295–3303
25. Huang L, Wang H, Wang Z, Mitra A, Bozhilov KN, Yan Y (2002) Nanowire arrays electrodeposited from liquid crystalline phases. Adv Mater 14:61–64
26. Sun Y, Mayers B, Xia Y (2003) Metal nanostructures with hollow interiors. Adv Mater 15:641–646
27. Mayers B, Jiang X, Sunderland D, Cattle B, Xia Y (2003) Hollow nanostructures of platinum with controllable dimensions can be synthesized by templating against selenium nanowires and colloids. J Am Chem Soc 125:13364–13365
28. Song J, Wu Y, Messer B, Kind H, Yang P (2001) Metal nanowire formation using Mo3Se3-as reducing and sacrificing templates. J Am Chem Soc 123:10397–10398
29. Yu J-S, Kim JY, Lee S, Mbindyo JK, Martin BR, Mallouk TE (2000) Template synthesis of polymer-insulated colloidal gold nanowires with reactive ends. Chem Commun 24:2445–2446
30. Payne EK, Shuford KL, Park S, Schatz GC, Mirkin CA (2006) Multipole plasmon resonances in gold nanorods. J Phys Chem B 110:2150–2154
31. Wang J (2008) Barcoded metal nanowires. J Mater Chem 18:4017–4020
32. Wirtz M, Martin CR (2003) Template-fabricated gold nanowires and nanotubes. Adv Mater 15:455–458
33. Sperling RA, Pellegrino T, Li JK, Chang WH, Parak WJ (2006) Electrophoretic separation of nanoparticles with a discrete number of functional groups. Adv Func Mater 16:943–948
34. Jin R, Cao YC, Hao E, Métraux GS, Schatz GC, Mirkin CA (2003) Controlling anisotropic nanoparticle growth through plasmon excitation. Nat 425–487
35. Wu X, Redmond PL, Liu H, Chen Y, Steigerwald M, Brus L (2008) Photovoltage mechanism for room light conversion of citrate stabilized silver nanocrystal seeds to large nanoprisms. J Am Chem Soc 130:9500–9506

36. Yang S, Zhang R, Wang Q, Ding B, Wang Y (2007) Coral-shaped 3D assemblies of gold nuclei induced by UV irradiation and its disintegration. Colloids Surf, A 311:174–179
37. Sau TK, Murphy CJ (2004) Room temperature, high-yield synthesis of multiple shapes of gold nanoparticles in aqueous solution. J Am Chem Soc 126: 8648–8649
38. Habas SE, Lee H, Radmilovic V, Somorjai GA, Yang P (2007) Shaping binary metal nanocrystals through epitaxial seeded growth. Nat Mater 6:692–697
39. Gole A, Murphy CJ (2004) Seed-mediated synthesis of gold nanorods: role of the size and nature of the seed. Chem Mater 16:3633–3640
40. Liu M, Guyot-Sionnest P (2005) Mechanism of silver (I)-assisted growth of gold nanorods and bipyramids. J Phys Chem B 109:22192–22200
41. Wirtz M, Martin CR (2003) Template-fabricated gold nanowires and nanotubes. Adv Mater 15(5):455–458
42. Preston CK, Moskovits M (1993) Optical characterization of anodic aluminum oxide films containing electrochemically deposited metal particles. 1. Gold in phosphoric acid anodic aluminum oxide films. J Phys Chem 97:8495–8503
43. Yu Y-Y, Chang S-S, Lee C-L, Wang CC (1997) Gold nanorods: electrochem ical synthesis and optical properties. J Phys Chem B 101:6661–6664
44. Yao H, Onishi T, Sato S, Kimura K (2002) High aspect ratio gold nanorods grown normal to high-energy surfaces. Chem Lett 31:458–459
45. Zhu J, Jin X-L (2007) Electrochemical synthesis of gold triangular nanoplates and self-organized into rhombic nanostructures. Superlattices Microstruct 41:271–276
46. Suslick KS, Choe S-B, Cichowlas AA, Grinstaff MW (1991) Sonochemical synthesis of amorphous iron. Nature 353:414
47. Suslick KS (1990) Sonochemistry. Science 247:1439–1445
48. Suslick KS, Flannigan DJ (2008) Inside a collapsing bubble: sonoluminescence and the conditions during cavitation. Annu Rev Phys Chem 59:659–683
49. Caruso RA, Ashokkumar M, Grieser F (2002) Sonochemical formation of gold sols. Langmuir 18:7831–7836
50. Mizukoshi Y, Fujimoto T, Nagata Y, Oshima R, Maeda Y (2000) Characterization and catalytic activity of core–shell structured gold/palladium bimetallic nanoparticles synthesized by the sonochemical method. J Phys Chem B 104:6028–6032
51. Gedanken A (2004) Using sonochemistry for the fabrication of nanomaterials. Ultrason Sonochem 11:47–55
52. Vinodgopal K, He Y, Ashokkumar M, Grieser F (2006) Sonochemically prepared platinum–ruthenium bimetallic nanoparticles. J Phys Chem B 110:3849–3852
53. Pol VG, Srivastava DN, Palchik O, Palchik V, Slifkin MA, Weiss AM, Gedanken A (2002) Sonochemical deposition of silver nanoparticles on silica spheres. Langmuir 18:3352–3357
54. Salkar R, Jeevanandam P, Aruna S, Koltypin Y, Gedanken A (1999) The sonochemical preparation of amorphous silver nanoparticles. J Mater Chem 9:1333–1335
55. Hadad L, Perkas N, Gofer Y, Calderon-Moreno J, Ghule A, Gedanken A (2007) Sonochemical deposition of silver nanoparticles on wool fibers. J Appl Polym Sci 104:1732–1737
56. Zhang J, Du J, Han B, Liu Z, Jiang T, Zhang Z (2006) Sonochemical formation of single-crystalline gold nanobelts. Angew Chem Int Ed 45:1116–1119
57. Mizukoshi Y, Okitsu K, Maeda Y, Yamamoto TA, Oshima R, Nagata Y (1997) Sonochemical preparation of bimetallic nanoparticles of gold/palladium in aqueous solution. J Phys Chem B 101:7033–7037
58. Anandan S, Grieser F, Ashokkumar M (2008) Sonochemical synthesis of Au–Ag core–shell bimetallic nanoparticles. J Phys Chem C 112:15102–15105
59. Okitsu K, Sharyo K, Nishimura R (2009) One-pot synthesis of gold nanorods by ultrasonic irradiation: the effect of pH on the shape of the gold nanorods and nanoparticles. Langmuir 25:7786–7790
60. Sánchez-Iglesias A, Pastoriza-Santos I, Pérez-Juste J, Rodríguez-González B, Garcia de Abajo FJ, Liz-Marzán LM (2006) Synthesis and optical properties of gold nanodecahedra with size control. Adv Mater 18:2529–2534

61. Jiang LP, Xu S, Zhu JM, Zhang JR, Zhu JJ, Chen HY (2004) Ultrasonic- assisted synthesis of monodisperse single-crystalline silver nanoplates and gold nanorings. Inorg Chem 43:5877–5883
62. Wiley B, Sun Y, Xia Y (2007) Synthesis of silver nanostructures with controlled shapes and properties. Acc Chem Res 40:1067–1076
63. Polshettiwar V, Nadagouda MN, Varma RS (2009) Microwave-assisted chemistry: a rapid and sustainable route to synthesis of organics and nanomaterials. Aust J Chem 62:16–26
64. Sreeram K, Nidhin M, Nair B (2008) Microwave assisted template synthesis of silver nanoparticles. Bull Mater Sci 31:937–942
65. Saifuddin N, Wong C, Yasumira A (2009) Rapid biosynthesis of silver nanoparticles using culture supernatant of bacteria with microwave irradiation. J Chem 6:61–70
66. Gou L, Chipara M, Zaleski JM (2007) Convenient, rapid synthesis of Ag nanowires. Chem Mater 19:1755–1760
67. Komarneni S, Li D, Newalkar B, Katsuki H, Bhalla AS (2002) Microwave- polyol process for Pt and Ag nanoparticles. Langmuir 18:5959–5962
68. Gao F, Lu Q, Komarneni S (2005) Interface reaction for the self-assembly of silver nanocrystals under microwave-assisted solvothermal conditions. Chem Mater 17:856–860
69. Yin H, Yamamoto T, Wada Y, Yanagida S (2004) Large-scale and size-controlled synthesis of silver nanoparticles under microwave irradiation. Mater Chem Phys 83:66–70
70. Abdelsayed V, Aljarash A, El-Shall MS, Al Othman ZA, Alghamdi AH (2009) Microwave synthesis of bimetallic nanoalloys and CO oxidation on ceria- supported nanoalloys. Chem Mater 21:2825–2834
71. Liu F-K, Huang P-W, Chang Y-C, Ko F-H, Chu T-C (2005) Combining optical lithography with rapid microwave heating for the selective growth of Au/Ag bimetallic core/shell structures on patterned silicon wafers. Langmuir 21:2519–2525
72. Liang X, Wang Z-J, Liu C-J (2010) Size-controlled synthesis of colloidal gold nanoparticles at room temperature under the influence of glow discharge. Nanoscale Res Lett 5:124–129
73. Hu B, Wang S-B, Wang K, Zhang M, Yu S-H (2008) Microwave-assisted rapid facile "green" synthesis of uniform silver nanoparticles: self-assembly into multilayered films and their optical properties. J Phys Chem C 112:11169–11174
74. Wang Z (2000) Transmission electron microscopy of shape-controlled nanocrystals and their assemblies. J Phys Chem B 104:1153–1175
75. Smith DJ, Bovin J, Bursill L, Petford-Long A, Ye H (1987) High resolution electron microscopy of surfaces and surface reactions. Surf Interface Anal 10:135–141
76. Li ZY, Young NP, Di Vece M, Palomba S, Palmer RE, Bleloch AL, Curley BC, Johnston RL, Jiang J, Yuan J (2008) Three-dimensional atomic- scale structure of size-selected gold nanoclusters. Nature 451:46–48
77. Huey BD, Sigmund W (2007) The application of scanning probe microscopy in materials science studies. JOM 59:11
78. Borchert H, Shevchenko EV, Robert A, Mekis I, Kornowski A, Grübel G, Weller H (2005) Determination of nanocrystal sizes: a comparison of TEM, SAXS, and XRD studies of highly monodisperse CoPt3 particles. Langmui 21:1931–1936
79. Kontoyannis CG, Vagenas NV (2000) Calcium carbonate phase analysis using XRD and FT-Raman spectroscopy. Analyst 125:251–255
80. Thamaphat K, Limsuwan P, Ngotawornchai B (2008) Phase characterization of TiO_2 powder by XRD and TEM. Kasetsart J (Nat Sci) 42:357–361
81. Corcoran C, Tavassol H, Rigsby M, Bagus P, Wieckowski A (2010) Application of XPS to study electrocatalysts for fuel cells. J Power Sources 195:7856–7879
82. Rodriguez JA, Goodman DW (1992) The nature of the metal-metal bond in bimetallic surfaces. Science 257:897–903
83. Richter B, Kuhlenbeck H, Freund H-J, Bagus PS (2004) Cluster core-level binding-energy shifts: the role of lattice strain. Phys Rev Lett 93:026805–026808
84. Yao Y, He DS, Lin Y, Feng X, Wang X, Yin P, Hong X, Zhou G, Wu Y, Li Y (2016) Modulating fcc and hcp ruthenium on the surface of palladium–copper alloy through tunable lattice mismatch. Angew Chem Int Ed 55:5501–5505

85. Spiro M, de Jesus DM (2000) Nanoparticle catalysis in microemulsions: oxidation of n, n-dimethyl-p-phenylenediamine by cobalt (III) pentaammine chloride catalyzed by colloidal palladium in water/AOT/n-heptane microemulsions. Langmuir 16:2464–2468
86. Scheuermann GM, Rumi L, Steurer P, Bannwarth W, Mülhaupt R (2009) Palladium nanoparticles on graphite oxide and its functionalized graphene derivatives as highly active catalysts for the Suzuki– Miyaura coupling reaction. J Am Chem Soc 131:8262–8270
87. Gensch T, Hopkinson M, Glorius F, Wencel-Delord J (2016) Mild metal- catalyzed C-H activation: examples and concepts. Chem Soc Rev 45:2900–2936
88. Narayanan R, El-Sayed MA (2004) Changing catalytic activity during colloidal platinum nanocatalysis due to shape changes: electron-transfer reaction. J Am Chem Soc 126:7194–7195
89. Narayanan R, El-Sayed MA (2003) Effect of catalysis on the stability of metallic nanoparticles: Suzuki reaction catalyzed by PVP-palladium nanoparticles. J Am Chem Soc 125:8340–8347
90. Narayanan R, El-Sayed MA (2005) Catalysis with transition metal nanoparticles in colloidal solution: nanoparticle shape dependence and stability. J Phys Chem B 109:12663–12676
91. Shiraishi Y, Toshima N (1999) Colloidal silver catalysts for oxidation of ethylene. J Mol Catal A: Chem 141:187–192
92. Falicov L, Somorjai G (1985) Correlation between catalytic activity and bonding and coordination number of atoms and molecules on transition metal surfaces: Theory and experimental evidence. Proc Natl Acad Sci 82:2207–2221
93. Mistry H, Varela AS, Kuehl S, Strasser P, Cuenya BR (2016) Nanostructured electrocatalysts with tunable activity and selectivity. Nat Rev Mater 1:16009–16022
94. Wu Y, Wang D, Li Y (2014) Nanocrystals from solutions: catalysts. Chem Soc Rev 43:2112–2124
95. He D, Zhang L, He D, Zhou G, Lin Y, Deng Z, Hong X, Wu Y et al (2016) Amorphous nickel boride membrane on a platinum–nickel alloy surface for enhanced oxygen reduction reaction. Nat Commun 7:12362–12369
96. He DS, He D, Wang J, Lin Y, Yin P, Hong X, Wu Y, Li Y (2016) Ultrathin cosahedral Pt-enriched nanocage with excellent oxygen reduction reaction activity. J Am Chem Soc 138:1494–1497
97. Zhu W, Michalsky R, Metin O, Lv H, Guo S, Wright CJ, Sun X, Peterson AA, Sun S (2013) Monodisperse Au nanoparticles for selective electro-catalytic reduction of CO_2 to CO. J Am Chem Soc 135:16833–16836
98. Zhu W, Zhang YJ, Zhang H, Lv H, Li Q, Michalsky R, Peterson AA, Sun S (2014) Active and selective conversion of CO_2 to CO on ultrathin Au nanowires. J Am Chem Soc 136:16132–16135
99. Mistry H, Reske R, Zeng Z, Zhao ZJ, Greeley J, Strasser P, Cuenya BR (2014) Exceptional size-dependent activity enhancement in the electroreduction of CO_2 over Au nanoparticles. J Am Chem Soc 136:16473–16476
100. Gloag L, Benedetti TM, Cheong S, Marjo CE, Gooding JJ, Tilley RD (2018) Cubic-core hexagonal-branch mechanism to synthesize bimetallic branched and faceted Pd-Ru nanoparticles for oxygen evolution reaction electrocatalysis. J Am Chem Soc 140(40):12760–12764
101. Chao T, Luo X, Chen W, Jiang B, Ge J, Lin Y, Wu G, Wang X, Hu Y, Zhuang Z, Wu Y, Hong X, Li Y (2017) Atomically dispersed copper– platinum dual sites alloyed with palladium nanorings catalyze the hydrogen evolution reaction. Angew Chem 129:16263–16267
102. Yang Y, Matsubara S, Xiong L, Hayakawa T, Nogami M (2007) Solvothermal synthesis of multiple shapes of silver nanoparticles and their SERS properties. J Phys Chem C 111(26):9095–9104
103. Ivy J (2004) Summary of electrolytic hydrogen production: milestone completion report. National Renewable Energy Lab, Golden, CO (US)
104. Mette K, Bergmann A, Tessonnier JP, Hävecker M, Yao L, Ressler T, Schlögl R, Strasser P, Behrens M (2012) Nanostructured manganese oxide supported on carbon nanotubes for electrocatalytic water splitting. ChemCatChem 4:851–862

105. Manabe A, Kashiwase M, Hashimoto T, Hayashida T, Kato A, Hirao K, Shimomura I, Nagashima I (2013) Basic study of alkaline water electrolysis. Electrochim Acta 100:249–256
106. Carmo M, Fritz DL, Mergel J, Stolten D (2013) A comprehensive review on PEM water electrolysis. Int J Hydrogen Energy 38:4901–4934
107. Ursua A, Gandia LM, Sanchis P (2012) Hydrogen production from water electrolysis: current status and future trends. Proc IEEE 100:410–426
108. Dau H, Limberg C, Reier T, Risch M, Roggan S, Strasser P (2010) The mechanism of water oxidation: from electrolysis via homogeneous to biological catalysis. ChemCatChem 2:724–761
109. Trasatti S (1984) Electrocatalysis in the anodic evolution of oxygen and chlorine. Electrochim Acta 29:1503–1512
110. Katsounaros I, Cherevko S, Zeradjanin AR, Mayrhofer KJ (2014) Oxygen electrochemistry as a cornerstone for sustainable energy conversion. Angew Chem Int Ed 53:102–121
111. Zhao S, Wang Y, Dong J, He CT, Yin H, An P, Zhao K, Zhang X, Gao C, Zhang L, Lv J, Wang J, Zhang J, Khattak AM, Khan NA, Wei Z, Zhang J, Liu S, Zhao H, Tang Z (2016) Ultrathin metal–organic framework nanosheets for electrocatalytic oxygen evolution. Nat Energy 1:16184–16193
112. Moreno-Hernandez IA, MacFarland CA, Read CG, Papadantonakis KM, Brunschwig BS, Lewis NS (2017) Crystalline nickel manganese antimonate as a stable water-oxidation catalyst in aqueous 1.0 M H2SO4. Energy Environ Sci 10:2103–2108
113. Danilovic N, Subbaraman R, Chang KC, Chang SH, Kang YJ, Snyder J, Paulikas AP, Strmcnik D, Kim Y-T, Myers D, Stamenkovic VR, Markovic NM (2014) Activity–stability trends for the oxygen evolution reaction on monometallic oxides in acidic environments. J Phys Chem Lett 5:2474–2478
114. Antolini E (2014) Iridium as catalyst and cocatalyst for oxygen evolution/reduction in acidic polymer electrolyte membrane electrolyzers and fuel cells. ACS Catal 4:1426–1440
115. Cherevko S, Zeradjanin AR, Topalov AA, Kulyk N, Katsounaros I, Mayrhofer KJ (2014) Dissolution of noble metals during oxygen evolution in acidic media. ChemCatChem 6:2219–2223
116. Cherevko S, Geiger S, Kasian O, Kulyk N, Grote JP, Savan A, Shrestha BR, Merzlikin S, Breitbach B, Ludwig A, Mayrhofer KJ (2016) Oxygen and hydrogen evolution reactions on Ru, RuO_2, Ir, and IrO_2 thin film electrodes in acidic and alkaline electrolytes: a comparative study on activity and stability. Catal Today 262:170–180
117. Mozota J, Conway B (1983) Surface and bulk processes at oxidized iridium electrodes—I. Monolayer stage and transition to reversible multilayer oxide film behaviour. Electrochimica Acta 28:1–8
118. Conway B, Mozota J (1983) Surface and bulk processes at oxidized iridium electrodes—II. Conductivity-switched behaviour of thick oxide films. Electro chimica Acta 28:9–16
119. Frydendal R, Paoli EA, Knudsen BP, Wickman B, Malacrida P, Stephens IE, Chorkendorff I (2014) Benchmarking the stability of oxygen evolution reaction catalysts: the importance of monitoring mass losses. ChemElectroChem 1:2075–2081
120. Matsumoto Y, Tazawa T, Muroi N, Sato E (1986) New types of anodes for the oxygen evolution reaction in acidic solution. J Electrochem Soc 133:2257–2262
121. Exner KS, Anton J, Jacob T, Over H (2015) Ligand effects and their impact on electrocatalytic processes exemplified with the oxygen evolution reaction (OER) on RuO_2 (110). ChemElectroChem 2:707–713
122. Audichon T, Mayousse E, Morisset S, Morais C, Comminges C, Napporn TW, Kokoh KB (2014) Electroactivity of RuO_2-IrO_2 mixed nanocatalysts toward the oxygen evolution reaction in a water electrolyzer supplied by a solar profile. Int J Hydrogen Energy 39:16785–16796
123. Mamaca N, Mayousse E, Arrii-Clacens S, Napporn TW, Servat K, Guillet N, Kokoh KB (2012) Electrochemical activity of ruthenium and iridium based catalysts for oxygen evolution reaction. Appl Catal B 111:376 380

124. Reier T, Pawolek Z, Cherevko S, Bruns M, Jones T, Teschner D, Selve S, Bergmann A, Nong HN, Schlögl R, Mayrhofer KJJ, Strasser P (2015) Molecular insight in structure and activity of highly efficient, low-Ir Ir–Ni oxide catalysts for electrochemical water splitting (OER). J Am Chem Soc 137:13031–13040
125. Chen Z, Zhao H, Zhang J, Xu J (2017) IrNi nanoparticle-decorated flower- shaped $NiCo_2O_4$ nanostructures: controllable synthesis and enhanced electrochemical activity for oxygen evolution reaction. Sci China Mater 60:119–130
126. Nong HN, Oh HS, Reier T, Willinger E, Willinger MG, Petkov V, Teschner D, Strasser P (2015) Oxide-supported IrNiOx core-shell particles as efficient, cost-effective, and stable catalysts for electrochemical water splitting. Angew Chem Int Ed 54:2975–2979
127. Petrykin V, Bastl Z, Franc J, Macounova K, Makarova M, Mukerjee S, Ramaswamy N, Spirovova I, Krtil P (2009) Local structure of nanocrystalline Ru1−xNixO2−δ dioxide and its implications for electrocatalytic behavior-An XPS and XAS Study. J Phys Chem C 113:21657–21666
128. Nong HN, Gan L, Willinger E, Teschner D, Strasser P (2014) IrOx core- shell nanocatalysts for cost-and energy-efficient electrochemical water splitting. Chem Sci 5:2955–2963
129. Abreu-Sepulveda M, Trinh P, Malkhandi S, Narayanan SR, Jorné J, Quesnel DJ, Postonr Jr JA, Manivannan A (2015) Investigation of oxygen evolution reaction at $LaRuO_3$, La3.5Ru4O13, and La2RuO5. Electrochimica Acta 180:401–408
130. Audichon T, Morisset S, Napporn TW, Kokoh KB, Comminges C, Morais C (2015) Effect of adding CeO_2 to RuO_2-IrO_2 mixed nanocatalysts: activity towards the oxygen evolution reaction and stability in acidic media. ChemElectroChem 2:1128–1137
131. Näslund LÅ, Sánchez-Sánchez CM, Ingason ÁS, Bäckström J, Herrero E, Rosen J, Holmin S (2013) The role of TiO_2 doping on RuO_2-coated electrodes for the water oxidation reaction. J Phys Chem C 117:6126–6135
132. Puthiyapura VK, Pasupathi S, Basu S, Wu X, Su H, Varagunapandiyan N, Pollet B, Scott K (2013) RuxNb1−xO2 catalyst for the oxygen evolution reaction in proton exchange membrane water electrolysers. Int J Hydrogen Energy 38:8605–8616
133. Musiani M, Furlanetto F, Bertoncello R (1999) Electrodeposited PbO_2+RuO_2: a composite anode for oxygen evolution from sulphuric acid solution. J Electroanal Chem 465:160–167
134. Wu X, Tayal J, Basu S, Scott K (2011) Nano-crystalline RuxSn1−xO2 powder catalysts for oxygen evolution reaction in proton exchange membrane water electrolysers. Int J Hydrogen Energy 36:14796–14804
135. Neyerlin K, Bugosh G, Forgie R, Liu Z, Strasser P (2009) Combinatorial study of high-surface-area binary and ternary electrocatalysts for the oxygen evolution reaction. J Electrochem Soc 156:B363–B369
136. De-Pauli C, Trasatti S (2002) Composite materials for electrocatalysis of O2 evolution: IrO2+ SnO2 in acid solution. J Electroanal Chem 538:145–151
137. Chen G, Chen X, Yue PL (2002) Electrochemical behavior of novel Ti/IrOx−Sb_2O_5−SnO_2 Anodes. J Phys Chem B 106:4364–4369
138. Tae EL, Song J, Lee AR, Kim CH, Yoon S, Hwang IC, Kim MG, Yoon KB (2015) Cobalt oxide electrode doped with iridium oxide as highly efficient water oxidation electrode. ACS Catal 5:5525–5529
139. Petrykin V, Macounova K, Shlyakhtin O, Krtil P (2010) Tailoring the selectivity for electrocatalytic oxygen evolution on ruthenium oxides by zinc substitution. Angew Chem 122:4923–4925
140. Petrykin V, Macounová K, Okube M, Mukerjee S, Krtil P (2013) Local structure of Co doped RuO_2 nanocrystalline electrocatalytic materials for chlorine and oxygen evolution. Catal Today 202:63–69
141. Wang C, Sui Y, Xiao G, Yang X, Wei Y, Zou G, Zou B (2015) Synthesis of Cu–Ir nanocages with enhanced electrocatalytic activity for the oxygen evolution reaction. J Mater Chem A 3:19669–19673
142. Wu X, Scott K (2011) RuO_2 supported on Sb-doped SnO_2 nanoparticles for polymer electrolyte membrane water electrolysers. Int J Hydrogen Energy 36:5806–5810

143. Zou X, Zhang Y (2015) Noble metal-free hydrogen evolution catalysts for water splitting. Chem Soc Rev 44:5148–5180
144. Quaino P, Juarez F, Santos E, Schmickler W (2014) Volcano plots in hydrogen electrocatalysis–uses and abuses. Beilstein J Nanotechnol 5:846–854
145. Morales-Guio CG, Stern L-A, Hu X (2014) Nanostructured hydrotreating catalysts for electrochemical hydrogen evolution. Chem Soc Rev 43:6555–6569

Chapter 2
Modulating FCC and HCP Ruthenium on the Surface of Palladium–Copper Alloy Through Tunable Lattice Mismatch

2.1 Introduction

Metal nanoparticles (NPs) are emerging as a central nanomaterial for catalysis [1], plasmonic [2], sensing [3], and so on. Many of them, such as Fe [4], Ni [5, 6], Ag [7], Au [8, 9] and Ru [10] NPs can exhibit multiple crystallographic structures (e.g. face centered cubic (fcc), hexagonal close packed (hcp), and body centered cubic (bcc)) under different conditions. In addition to the well-studied features, such as size, composition, and shape, the crystallographic structure of metal NPs can also critically influence their functionalities [11–13]. Achieving the crystallographic structure control of metal NPs which will enable us to better explore their properties is highly desired for synthetic methodology studies. Traditionally in the bulk form, the structural control of a metal usually relies on altering the temperature and pressure [8]. Such rigid and expensive tuning methods are not suitable for practical applications. At the nanoscale, however, the surface compression or tensile stress may play a critical role in directing the crystallographic structure [14, 15]. For instance, driven by the minimization process of surface energy, solvent exchange will lead to the transformation from icosahedral Au_{13} clusters protected by dodecanethiol and triphenylphosphine to cuboctahedral structure protected by triphenylphosphine via structural rearrangement [15]. Though much effort has been made [7, 8], it still remains a great challenge to modulate the crystallographic structure of metal NPs through mild chemical methods.

Epitaxial growth is a versatile and facile approach to construct core–shell, yolk–shell, or hybrid structures by depositing a crystalline overlayer on a crystalline substrate [16–19]. Taking the substrate as a seed, the crystallographic structure of the overlayer usually preserves the same epitaxy with respect to the substrate in the case of moderate lattice mismatches between two materials. Otherwise, the overlayer is either differently orientated with respect to the substrate or randomly located on the surface of the substrate [20]. Under given temperature and pressure, the lattice mismatch between two materials is usually unchangeable. As such, modulating the epitaxial or non-epitaxial mode that the overlayer adopts to grow on the substrate

Y. Yao, *Controllable Synthesis and Atomic Scale Regulation of Noble Metal Catalysts*, Springer Theses, https://doi.org/10.1007/978-981-19-0205-5_2

cannot be implemented. An alloy is an exception, whereas its lattice parameter is tunable at some level by varying the composition ratio of two metals. This unique feature may impart flexibility to designing the lattice mismatch between the substrate and the overlayer and offer the ability to manipulate the growth mode of the overlayer.

Ruthenium is one of the most promising metal materials because of its importance in catalyzing a wide range of reactions, such as synthesis of ammonia [21], Fischer–Tropsch synthesis [21], and CO oxidation [22]/methanation [23]. Most recently, Joo and Kusada have discovered the existence of the nanoscale fcc Ru, which is not conventional and is not favored at ambient conditions, and reported its relative different catalytic performance with respect to the hcp Ru [10, 22]. To date, the synthetic methodology of controlling the crystallographic structure of Ru or its heterostructures is still scarce [24]. Herein, we demonstrate the epitaxial growth can be utilized to induce the fcc Ru overlayer on the Pd–Cu alloy surface by carefully tuning the composition of bimetallic alloy substrate. In addition, when the composition ratio of Pd and Cu was varied, the enlarged lattice mismatch could not be compensated and the epitaxial growth was broken. Thus, the hcp Ru, which is the conventionally stable structure [25], dominated at the overlayer. This is the first report that shows the modulating of the crystallographic structure of an overlayer on the same type of substrate through tunable lattice mismatch.

2.2 Experimental Section

2.2.1 Materials

Name	Formula	Purity	Manufacture
Ruthenium trichloride	$RuCl_3{\cdot}xH_2O$	A.R	Sigma-Aldrich
Palladium acetylacetonate	$Pd(acac)_2$	A.R	Alfa Aesar
Copper chloride	$CuCl_2{\cdot}2H_2O$	A.R	Sinopharm
Ethanol	CH_3OH	A.R	Sinopharm
Hexane	C_6H_{14}	A.R	Sinopharm
Oleylamine	OAm	A.R	Sigma-Aldrich
Octadecene	ODE	A.R	Sigma-Aldrich
Styrene	C_8H_8	98%	Sinopharm
4-nitrochlorobenzene	$C_6H_4ClNO_2$	98%	Sinopharm
Tolane	$C_{14}H_{10}$	98%	Energy chemical
Trans-stilbene	$C_{14}H_{12}$	98%	Energy chemical
Cis-stilbene	$C_{14}H_{12}$	98%	Energy chemical
diphenylethane	$C_{14}H_{14}$	98%	Sinopharm
5% Nafion	Nafion	5%	Sigma-Aldrich

(continued)

(continued)

Name	Formula	Purity	Manufacture
Nitrogen	N_2	99.99%	Nanjing Special Gas Corp
Hydrogen	H_2	99.99%	Nanjing Special Gas Corp

The reagents were directly used without purification.

2.2.2 *Methods*

Preparation of Pd–Cu nanocrystals. In a typical synthesis of Pd_xCu_{1-x} ($0 < x < 1$) nanocrystals, $Pd(acac)_2$ (7.5 mg), and $CuCl_2{\cdot}2H_2O$ (ranging from 0 to 40 mg), were dissolved in 1 mL ethanol and 3 ml oleylamine (OAM). The mixture was sonicated for 15 min. The as-obtained transparent dark blue solution was slowly injected to a vial containing 3 ml 1-octadecylene (ODE) which was preheated at 120 °C for 10 min with vigorously stirring. The reaction took 24 h. After being cooled down to room temperature, the products were precipitated by ethanol, separated via centrifugation and further purified by an ethanol-hexane mixture for three times.

Preparation of Pd–Cu@Ru yolk-shell nanocrystals. In a typical synthetic method to prepare Pd–Cu@Ru yolk-shell nanocrystals, 15.6 mg $RuCl_3{\cdot}xH_2O$ dissolving in 1 ml ethanol was added to the as-prepared Pd–Cu seed and then the temperature was changed from 120 to 200 °C, followed by 12 h thermal treating at 200 °C. Finally, the products were precipitated by ethanol, separated via centrifugation and further purified by an ethanol-hexane mixture.

Preparation of Ru hollow cage. In a typical synthetic method to Ru hollow cage, Pd–Cu@Ru yolk-shell nanocrystals were dissolved in 10 ml aqua regia, followed by 10 min stirring at room temperature. Finally, the products were separated via centrifugation and further purified by ethanol.

2.2.3 *Characterizations*

The crystallographic structure and phase purity were determined by Rigaku RU-200b X-ray powder diffractometer with CuKa radiation (l = 1.5418 Å). The inductively coupled plasma-mass spectrometry (ICP-MS) and energy dispersive spectrometer (EDS) were used to measure the composition of the product. Low magnification TEM images were recorded by a Hitachi-7650 working at 100 kV. High resolution HAADF-STEM image was collected on JEOL ARM 200F and JEM 2100 F with probe corrector operating at 200 kV.

2.2.4 *Electrochemical Measurements*

Preparation of working electrode. The Pd–Cu alloy seeds were loaded on carbon (wt 80%), followed by thermal treatment at 400 °C for 40 min. Then the product was dispersed in 1 ml ethanol, and 50 μl Nafion were added, followed by 10 min sonication. Then, 10 μl of the dispersion was transferred onto the glassy carbon electrode (GCE) by a pipette.

Potential measurements of Pd–Cu alloy electrode. The potential of Pd–Cu alloy working electrode were measured according to the procedure described by the reference, which made use of a cell accommodating the working electrode and the AgCl/Ag/KCl (saturated) electrode. The working electrode was made as described above. First of all, the AgCl/Ag/KCl (saturated) electrode was sealed through the bottom of a small glass tube. Then, carefully insert the working electrode into the cell, which previously has been filled with a 0.1 M $CuCl_2$ solution at room temperature. The electrochemical analyzer was used to measure voltage of the cell. All of the potentials of Pd–Cu working electrodes were converted to values versus standard hydrogen electrode according to the standard reversible AgCl/Ag/KCl (saturated) electrode (0.1981 V).

2.2.5 *Organic Catalysis*

Typical procedure for the catalytic hydrogenation of styrene. First, 20 μL styrene (0.17 mmol) in 1.5 mL toluene and the catalyst (contain 0.005 mmol Ru, 3 mol %) were added in a 10 mL round flask. Then, the round flask was purged with H_2 to completely remove air from the reactor and the reaction was allowed to proceed at 80 °C. The progress of the reaction was monitored by GC. Hydrogenation rates for styrene were calculated on the basis of the consumption rates for the substrates.

Typical procedure for the catalytic semihydrogenation of diphenylacetylene. First, 10 mg diphenylacetylene (0.056 mmol) in 1.5 mL toluene and the catalyst (contain 0.005 mmol Ru) were added in a 10 mL round flask. Then, the round flask was purged with H_2 to completely remove air from the reactor and the reaction was allowed to proceed at 80 °C. The progress of the reaction was monitored by GC. Hydrogenation rates for diphenylacetylene were calculated on the basis of the consumption rates for the substrates.

Typical procedure for the catalytic hydrogenation of 4-nitrochlorobenzene. First, 6 mg 4-nitrochlorobenzene dispersed in a mixed solvent containing 0.5 mL toluene and 1.5 ml DMF and the catalyst (contain 0.005 mmol Ru) were added in a 10 mL round flask. Then, the reaction mixture was stirred at 95 °C under H_2 balloon. The progress of the reaction was monitored by GC–MS.

2.3 Results and Discussion

Figure 2.1 illustrates the epitaxial growth of the fcc Ru shell on a Pd–Cu alloy seed and the shape transformation from Pd–Cu@Ru core–shell to yolk–shell architectures. The processes were traced by transmission electron microscopy (TEM) at three representative stages (0, 6, 12 h). We first adopted a solvothermal method to synthesize $PdCu_3$ alloy seeds with a homogeneous truncated octahedral shape and uniform size (19.6 ± 0.8 nm), as shown in Fig. 2.1a (the composition ratio was measured by inductively coupled plasma mass spectrometry (ICP-MS). The epitaxial growth was initially induced by the galvanic replacement between Ru and $PdCu_3$ seeds, during which the Pd–Cu seeds gradually adopted a Pd–Cu@Ru core–shell structure (Fig. 2.1b). Evidenced by X-ray diffraction (XRD) spectra, no peak appears that can be assigned to hcp Ru and the peaks belonging to fcc $PdCu_3$ remained roughly unshifted throughout the entire process (Fig. 2.2). ICP-MS measurement showed that the Pd/Cu ratio was retained at nearly 1:3. Therefore, it was concluded that the RuIII would gradually replace both the Pd^0 and Cu^0 during the evolution process (Figs. 2.3 and 2.4). As shown in Fig. 2.1c, the galvanic replacement finally resulted in a well-defined Pd–Cu@Ru yolk–shell structure. The detailed one-to-one correspondence between the TEM images at the three reaction stages and the projections of the 3D models are depicted in Fig. 2.1 according to our experimental results and hypotheses.

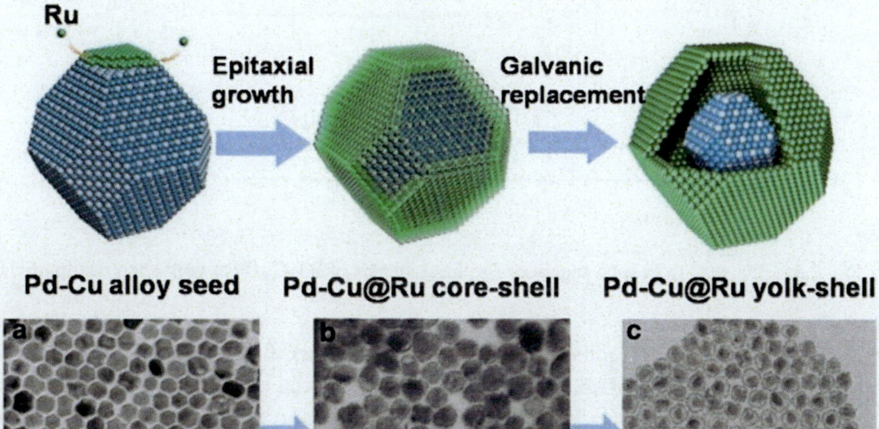

Fig. 2.1 Schematic illustrations and corresponding TEM images of the samples obtained at three representative stages during the structure evolution process. **a** $PdCu_3$ seed, **b** Pd–Cu@Ru core–shell NPs, and **c** Pd–Cu@Ru yolk–shell NPs

Fig. 2.2 XRD spectra as a function of the reaction time for the synthesis of Pd–Cu@Ru yolk-shell nanoparticles Pd–Cu@Ru

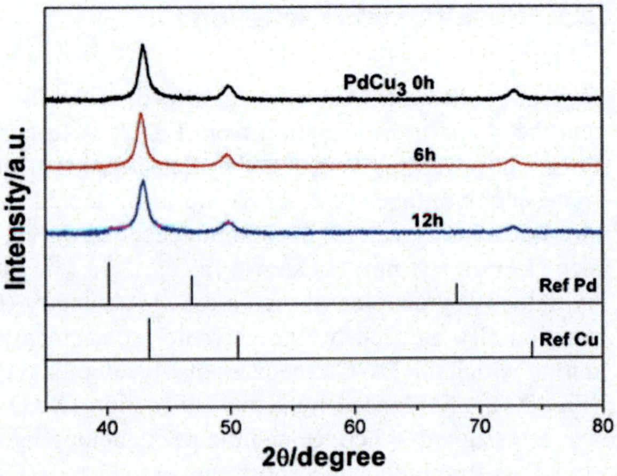

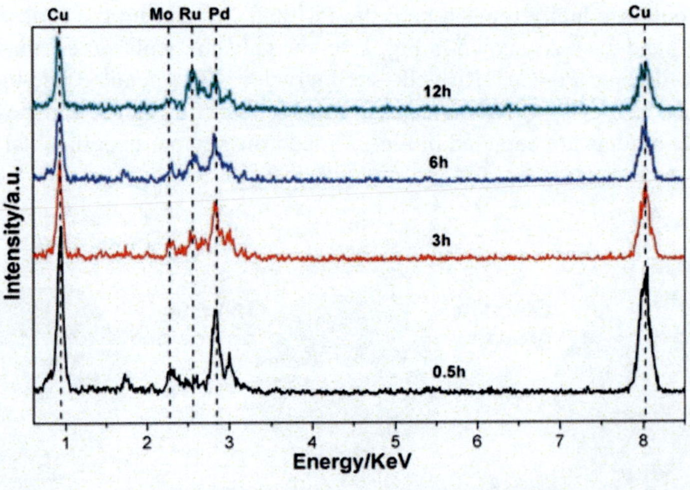

Fig. 2.3 EDS spectra in time sequence for the synthesis of Pd–Cu@Ru yolk-shell nanoparticles

The aberration-corrected high-angle annular dark-field scanning transmission electron microscopy (HAADF-STEM) images showed that all of the initial $PdCu_3$ seeds were fcc (Fig. 2.5). The corresponding elemental mappings revealed both Pd and Cu were uniformly distributed across the whole particle, in accordance with the nature of bimetallic alloy. The fast Fourier transforms (FFTs) ascribed to the atomic image of an individual $PdCu_3$ NP confirmed the standard fcc arrangement orientated along the [011] direction. According to the lattice spacing measurements in Fig. 2.5b, the average crystal constant of $PdCu_3$ (3.75 Å) was between that of Pd and Cu, which was consistent with the XRD measurement in Fig. 2.2.

The detailed electron micrograph of Pd–Cu@Ru core–shell structure after adding $RuCl_3$ solution is shown in Fig. 2.6. Indeed, the galvanic replacement between Ru

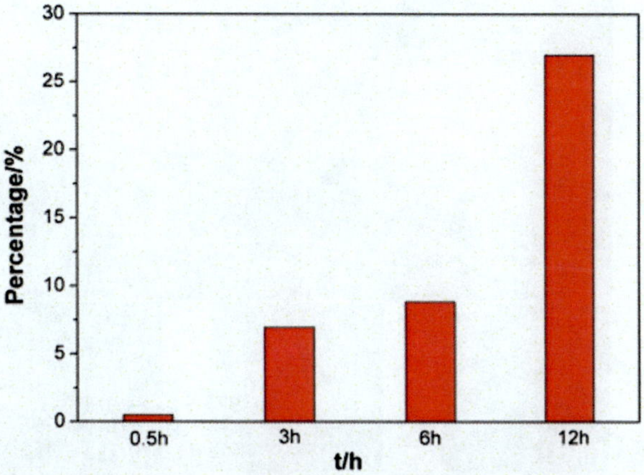

Fig. 2.4 The Ru percentage as a function of the reaction time (measured by EDS)

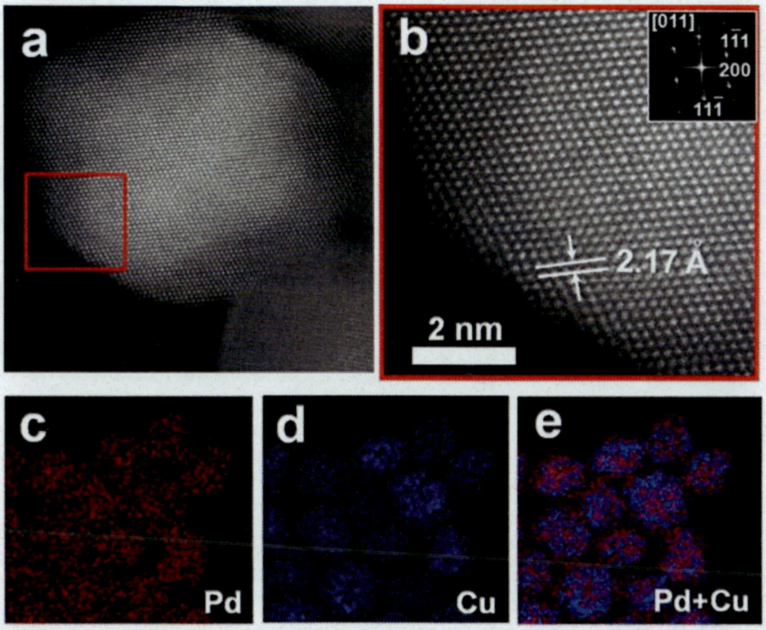

Fig. 2.5 HAADF-STEM images, the FFT pattern and EDS elemental mappings of $PdCu_3$ nanocrystals

and Pd was a slow process, which could be attributed to their roughly equivalent redox potentials. Alloying Pd with Cu would reduce its redox potential (Fig. 2.7), which was beneficial for the galvanic replacement between the $PdCu_3$ seed and Ru [26]. Surprisingly, the crystallographic structure of the Ru overlayer grown on the $PdCu_3$ surface was fcc phase as verified by the atomic HAADF-STEM images (Fig. 2.6a, b). Indicated by the white dashed line in Fig. 2.6b, it was very clear that the

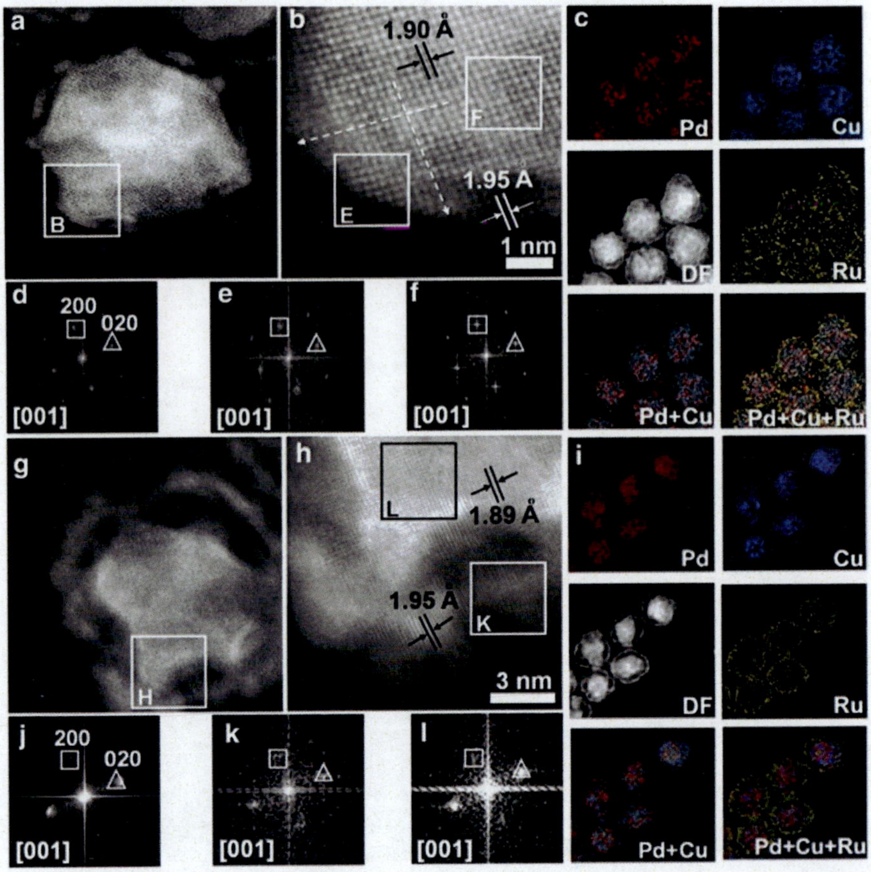

Fig. 2.6 **a**, **b** HAADF-STEM images, **c** EDS mappings, and **d–f** FFT patterns of Pd–Cu@Ru core–shell nanoparticles. **g**, **h** HAADF-STEM images, **i** EDS mappings, and **j–l** FFT patterns of Pd–Cu@Ru yolk–shell nanoparticles. See text for details. DF = dark field

Fig. 2.7 The measured electrical potentials of the as-prepared Pd–Cu alloy, Cu and Ru nanoparticles decorated working electrodes

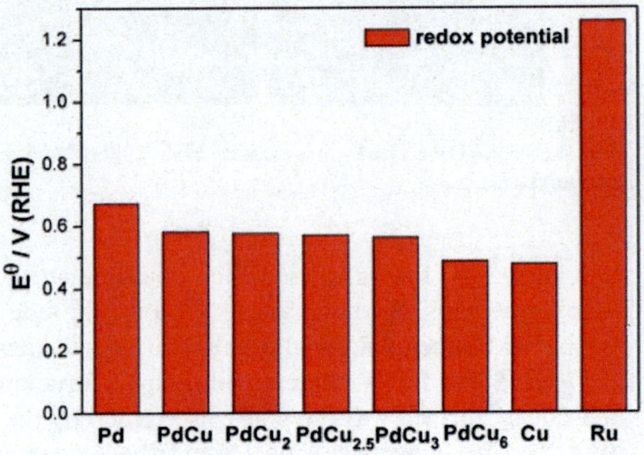

lattice of overgrown layers followed that of the core, showing the overgrowth of Ru was in a cubeon-cube epitaxial fashion. The FFT of an individual particle revealed the pattern (Fig. 2.6d) was a standard fcc pattern along with the [001] zone axis. The FFT patterns taken from the regions of core (Fig. 2.6f) and shell (Fig. 2.6e), respectively, further confirmed the atoms in these two areas adopt the same stacking mode. It was worth pointing out that the lattice spacing of core (1.90 Å) was close to that of shell (1.95 Å), allowing the epitaxial growth of Ru on the $PdCu_3$ surface. The corresponding elemental mappings clearly illustrated the core–shell structure, in which Pd and Cu concentrate at the center and Ru atoms located at the shell.

When extending the reaction to 12 h, most NPs evolved into the unique yolk–shell shape, with an average core size of 14.9 ± 1.4 nm and shell thickness of 2.9 ± 0.5 nm (Fig. 2.6g). The key in the formation of yolk–shell structure was the difference of the diffusion rate, which could be referred as Kirkendall effect [27]. Pd and Cu atoms diffused outwards faster than the inward diffusion of Ru atoms. This difference generates a net flux of vacancies from the surface to the center. The vacancies then coalesce into voids and develop preferably at the interface of the shell and core, which enables the formation of yolk–shell structure. Once the Ru atoms started to grow on the surface of $PdCu_3$, the newly deposited surface Ru atoms would serve as more tendentious deposition sites relative to $PdCu_3$ for further nucleation of Ru atoms. The HAADF-STEM image (Fig. 2.6h) of an individual NP suggested that Ru overlayer maintained the fcc atom packing throughout the structural evolution process, which is also supported by a sole set of diffraction patterns in the corresponding FFT pattern (Fig. 2.6j). To better strengthen this conclusion, we further compared two one-to-one FFTs of selected core and shell regions separately, thus reconfirming that the fcc overlayer can be induced by epitaxial growth on an fcc substrate (Fig. 2.6k, l). Carefully examinations at different areas confirmed the thickness of the Ru shells was mostly below 3 nm and the crystal phase adopts the fcc structure (Fig. 2.8). The corresponding elemental mappings were measured to clarify the spatial distribution of this trimetallic structure. Clearly shown in Fig. 2.6i, that the Pd–Cu alloy core is at the center of the hollow Ru shell with a clear gap between core and shell.

To reinforce that the fcc Pd–Cu alloy was encapsulated by the fcc Ru shell, we conducted an element-selective etching process to elucidate the real shape and crystallographic structure. In detail, the Ru was separated from the Pd and Cu through to its insolubility in aqua regia (HNO_3/HCl), leaving a residual hollow Ru cage (Fig. 2.9a). This process could be accurately traced by the line-scan profile in Fig. 2.10 and EDS spectra in Fig. 2.11, indicating the dissolution of the Pd–Cu core. The crystallographic structure of the hollow cage was identified by XRD(Fig. 2.9b) and high resolution HAADF-STEM images (Fig. 2.9c–e). The low-index XRD peaks revealed that the crystallographic structure of Ru hollow cage was fcc phase, which was in line with the previously discussion. The broadness of the peaks matched the HAADF-STEM observation of the hollow Ru NP in Fig. 2.9c very well, which shows that the Ru shell was as thin as 2–4 nm. The high resolution HAADF-STEM images (Fig. 2.9d, e) together with their FFTs from two selected areas also evidenced the characteristic fcc orientation along the [011] and [001] directions, respectively.

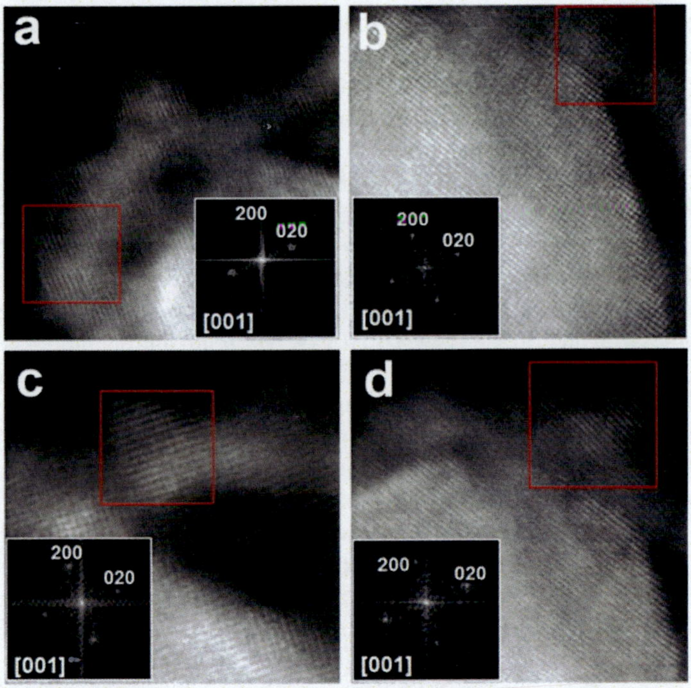

Fig. 2.8 HAADF-STEM images and the FFT patterns of $PdCu_3$@Ru nanoparticles

We further explored the crystallographic structure of grown Ru overlayer on a variety of comparable Pd–Cu alloy surface. They were synthesized using the same method with different ratios of Pd and Cu precursors. According to the Vegard's law [28], the lattice parameter of the Pd–Cu alloy would change correspondingly by varying the composition ratio of Pd and Cu, which gave the ability to tune the lattice spacing ranging from that of Pd to that of Cu. Figure 2.12 shows the TEM images of composition-dependent Pd–Cu alloy seeds with comparable morphology and size. The XRD patterns, as shown in Fig. 2.13, evidenced that the peaks indexed to {111}, {200}, and {220} diffractions of fcc structure situated between the standard fcc Pd and Cu and gradually left-shifted with the increasing Pd concentration. When Ru was deposited on those seeds, a different crystallographic structure of the Ru shell could be selectively obtained, whose XRD spectra (after treatment with aqua regia) are shown in Fig. 2.14a. Lattice mismatch is defined as the atomic distance ratio of Ru to Pd–Cu alloy. By building modes, the estimated lattice mismatch between (hcp and fcc) an Ru layer and Pd–Cu alloys is shown in Fig. 2.14b. In addition, we try to confirm the role of lattice mismatch on the structural preference of different ratio of Pd–Cu alloy. Pd–Cu seeds with a minor lattice mismatch with respect to fcc Ru, such as $PdCu_3$ and $PdCu_{2.5}$ were beneficial for the growth of a fcc Ru shell, in agreement with the HRTEM observations (Fig. 2.15). The deviation from the fcc Ru, no matter whether in higher or smaller concentration, such as Pd (Figs. 2.16 and 2.17), $PdCu_2$

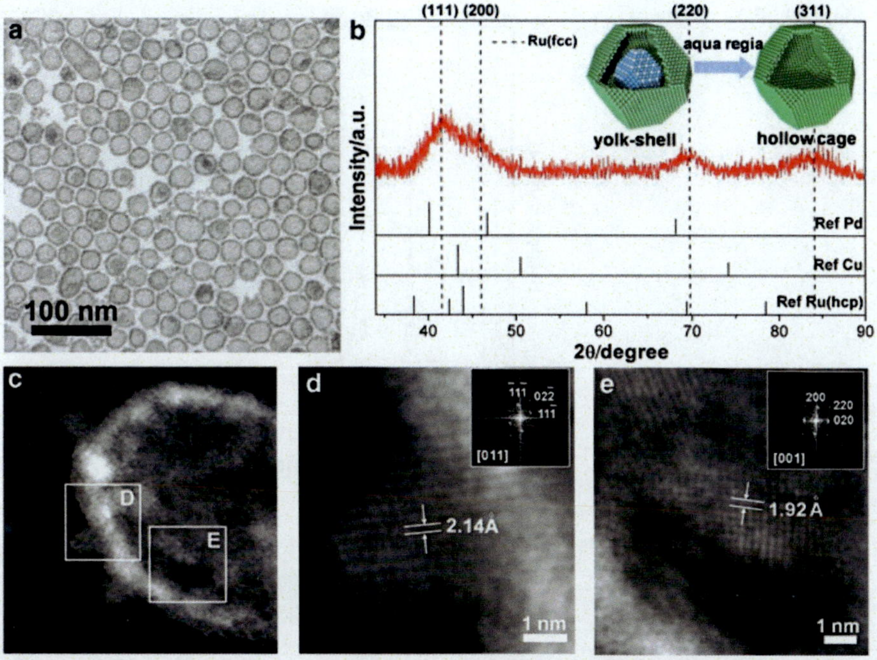

Fig. 2.9 **a** TEM image, **b** XRD pattern, and **c–e** HAADF-STEM images (insets are corresponding FFT patterns) of Ru hollow cage

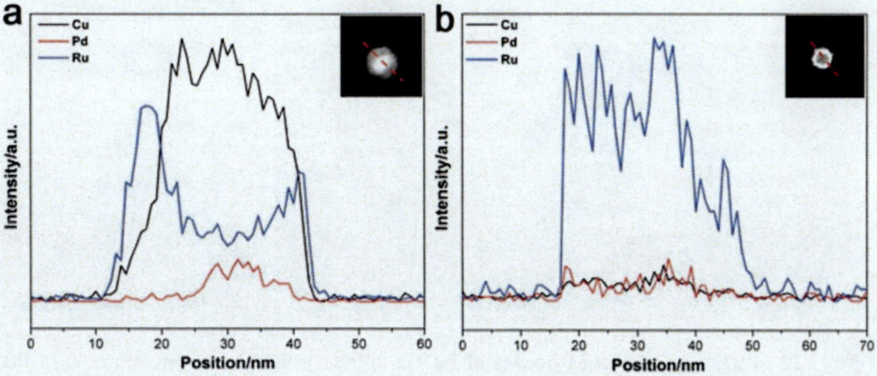

Fig. 2.10 EDS line profiles of **a** Pd–Cu@Ru yolk-shell nanoparticle, **b** Ru hollow cages obtained after treatment of aqua regia

(Fig. 2.18), or Cu (Figs. 2.19, 2.20 and 2.21), would drive the generation of the dominant hcp Ru. That is, once the lattice mismatch was increased to demolish the epitaxial growth, the Ru overlayer would transform to its conventional hcp structure.

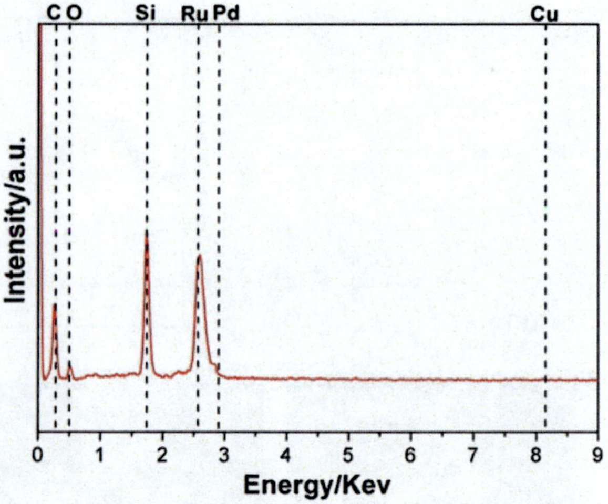

Fig. 2.11 EDS spectra of $PdCu_3$@Ru nanoparticles after treating with aqua regia

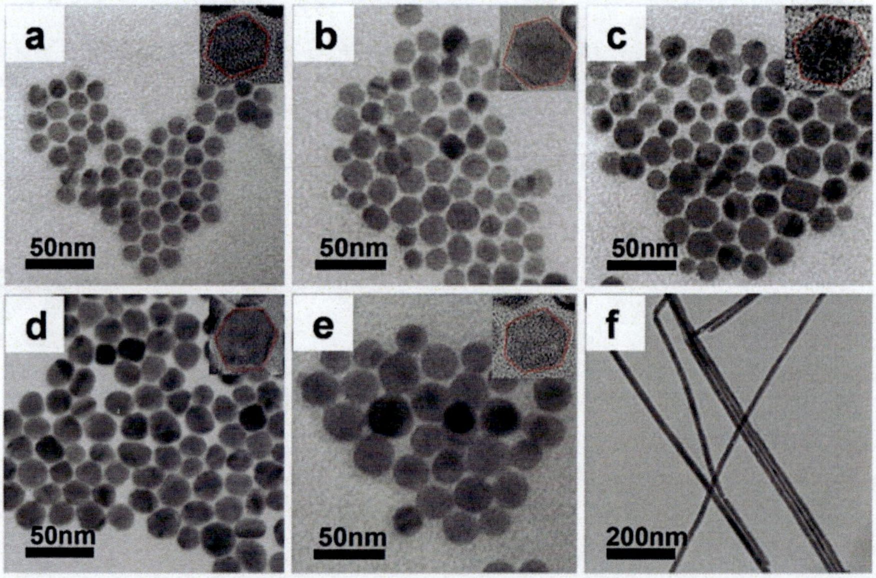

Fig. 2.12 **a** TEM and HRTEM images of Pd–Cu alloys by varying the atomic ratio of Pd/Cu (measured by ICP-MS). **a** Pd, **b** Pd_1Cu_1, **c** Pd_1Cu_2, **d** $Pd_1Cu_{2.5}$, **e** Pd_1Cu_6, **f** CuPd

DFT Calculations

From the viewpoint of energetics, the stability of the epitaxial film or the determination of growth modes could be evaluated by the binding energy of two parts, that is, the competition between chemical bonding energy, contributed by new bonds across the interface, and the elastic energy, caused by lattice mismatch [29, 30]. Density

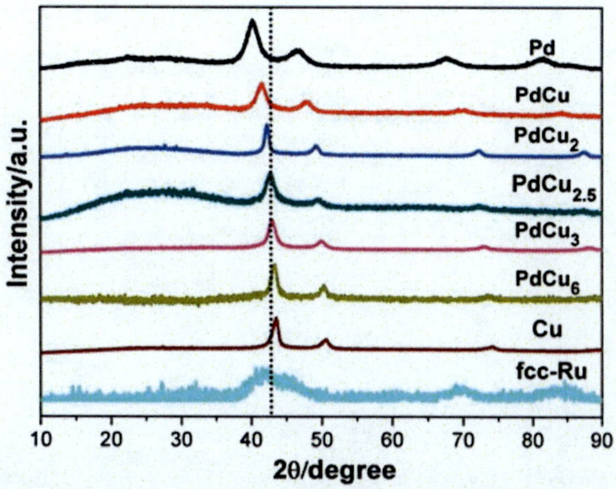

Fig. 2.13 XRD spectra of Pd–Cu alloy seeds with various Pd/Cu ratios and $PdCu_3$@Ru after treating with aqua regia

functional theory (DFT) calculations in the slab periodic model showed that the lattice constant of fcc Ru layers was 3.75 Å, very close to the experimental value (3.80 Å). The calculated binding energy of fcc Ru on the $PdCu_3$(001) surface (− 6.85 eV/Ru atom) is far stronger than those of fcc Ru and hcp Ru on $PdCu_3$(111) ones (− 4.34 and − 4.66 eV/Ru atom; Fig. 2.22). So in what follows, we focus on the situation of Ru on the exposed (001) facet of truncated octahedral $PdCu_3$, as verified by TEM (see Fig. 2.1). The lattice mismatch between fcc Ru(001) and $PdCu_3$-(001) was 0.54%, only producing an elastic energy of 0.11 eV. In contrast, the binding energy of fcc Ru(001) to $PdCu_3$(001) surfaces was estimated to be up to − 13.04 eV. In this manner, the contribution from the lattice mismatch was less important, and consequently, the chemical bonding energy dominated the nucleation behavior of Ru on the $PdCu_3$ substrate with epitaxial growth of Ru with the same structure as the $PdCu_3$ substrate. So the growth proceeded in a layer-by-layer fashion, just as in our experimental observations.

By comparison, the nucleation behavior of hcp Ru on a $PdCu_3$(001) substrate was more complicated because an ideal coherent interface was extremely difficult to achieve. To minimize the surface energy of the substrate, the nucleation in general required the reduced Ru atoms to bind to the surface atoms as much as possible, showing an ideal coherent interface [31]. This allows, to the greatest extent possible, the contacting hcp Ru to have the same number of and similar arrangement of atoms as the $PdCu_3$(001) surface, typically the (110) surface. The calculated surface lattice constants of hcp Ru(110) were 4.28 and 4.64 Å, much larger than that of $PdCu_3$-(001). More precisely, the lattice mismatches along the [100] and [010] directions of the $PdCu_3$(001) substrate were up to 14.6 and 24.4%, respectively. Our further dynamics simulations illustrated that upon the nucleation assumed in epitaxial growth, the induced elastic energy was estimated to reach 6.39 eV, and completely counteract the stacking energy difference of − 1.66 eV between the hcp Ru and fcc Ru structures. As a consequence, under induced stress conditions, the Ru atoms at the interface

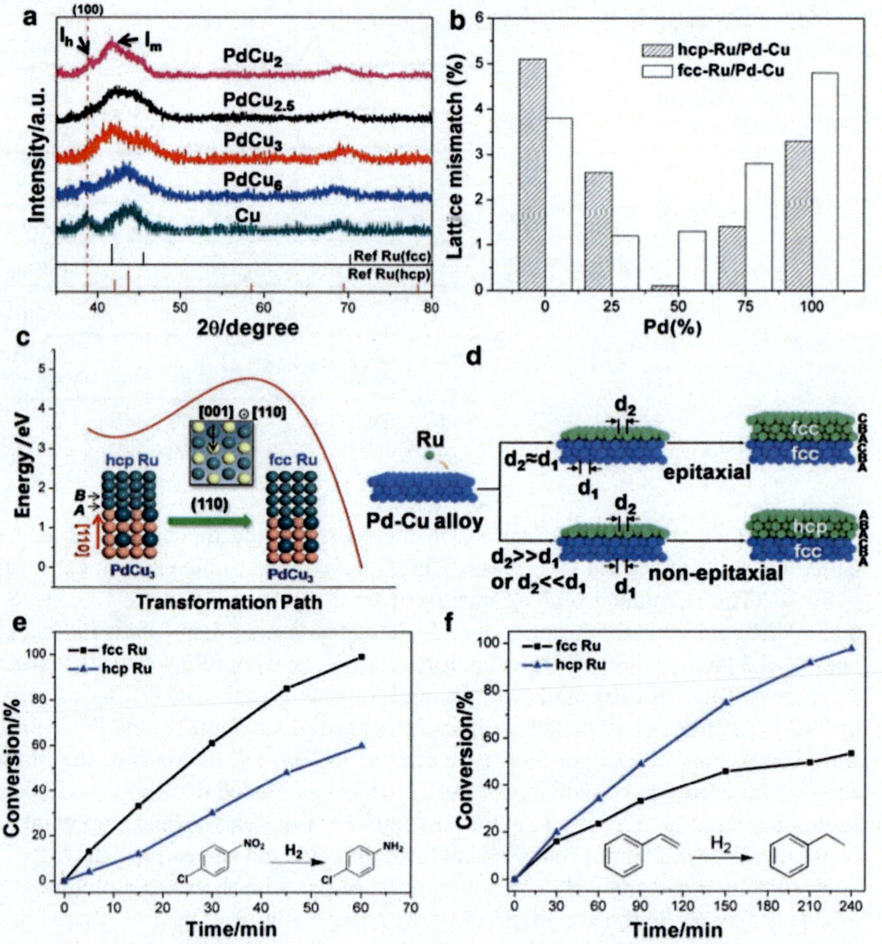

Fig. 2.14 **a** XRD spectra of Pd–Cu@Ru nanoparticles after treatment with aqua regia. The Pd/Cu ratios within the initial seed are 1/2, 1/2.5, 1/3, 1/6, and 0/1, respectively (From top to bottom). **b** Lattice mismatch between (hcp and fcc) Ru and Pd–Cu alloys with the content of Pd. **c** Structural transformation of hcp Ru to fcc Ru on the $PdCu_3$ substrate driven by the elastic energy, caused by lattice mismatch. Green, blue and red balls correspond to Ru, Pd, and Cu atoms, respectively. **d** Schematic illustrations of the epitaxial and non-epitaxial growth of Ru on a Pd–Cu surface. **e** Hydrogenation of 4-nitrochlorobenzene and **f** hydrogenation of styrene catalyzed by hcp Ru and fcc Ru

(A layer; green in top inset of Fig. 2.14c) moved along the [001] direction, and subsurface Ru atom (B layer, yellow in top inset of Fig. 2.14c) upshifted along the [110] direction. Finally, the hcp to fcc structure transformation was induced (Fig. 2.14c). Even so, the structural transformation, driven by thermodynamics as there was a rather large contribution of the lattice mismatch energy to the Gibbs free energy, was not spontaneous, and needed to overcome a barrier of at least 1.15 eV (the

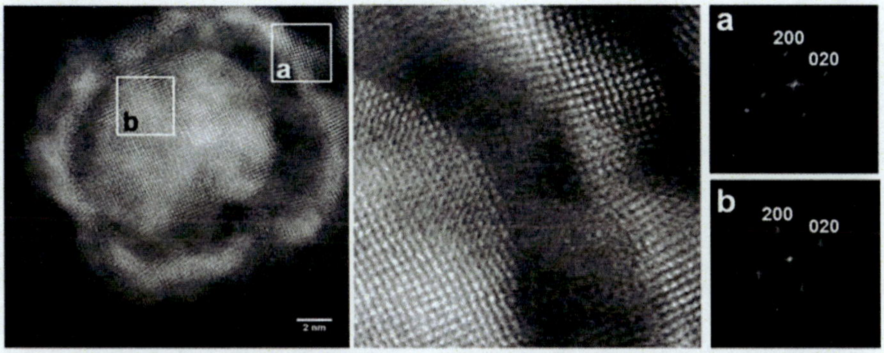

Fig. 2.15 HAADF-STEM images and the FFT patterns of $PdCu_{2.5}@Ru$ nanoparticles

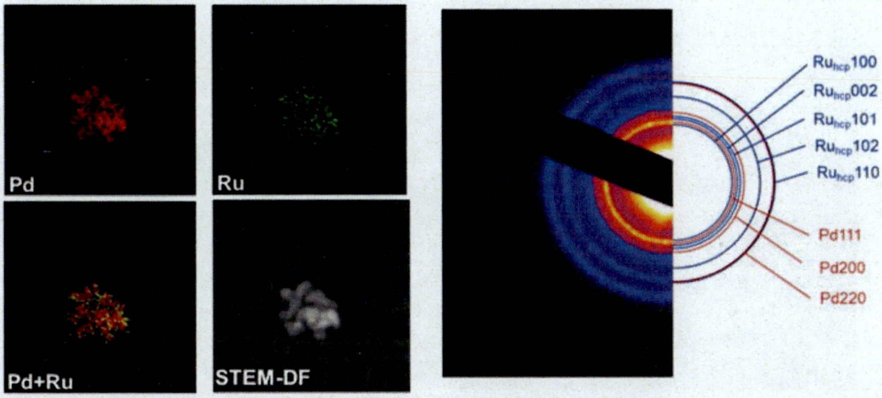

Fig. 2.16 EDS elemental mappings and the electron diffraction of Pd-Ru nanoparticles

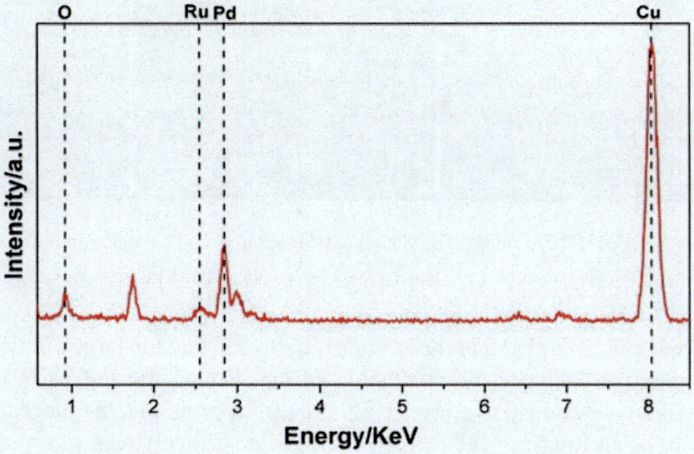

Fig. 2.17 EDS of Pd-Ru nanoparticles

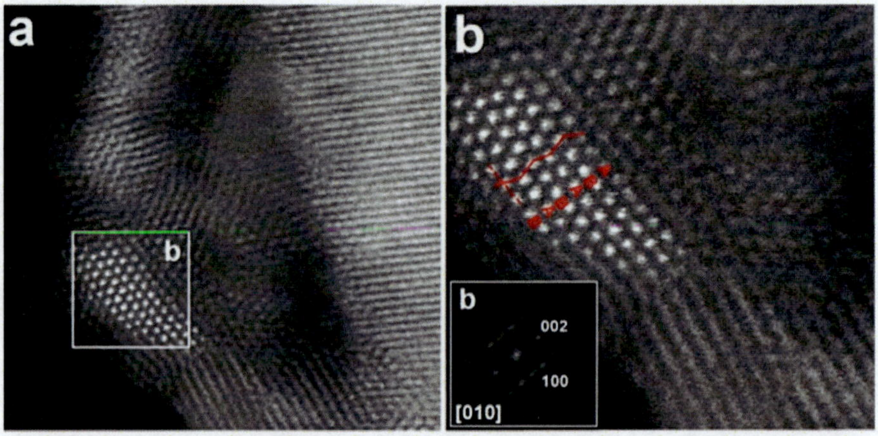

Fig. 2.18 HAADF-STEM images and the FFT pattern of $PdCu_2$@Ru nanoparticle

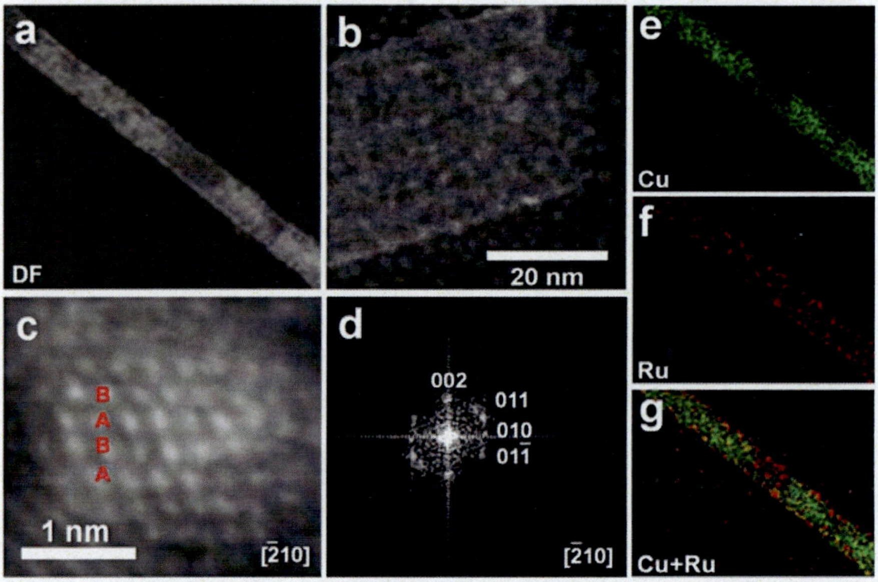

Fig. 2.19 HAADF-STEM images, EDS elemental mappings and the electron diffraction of Cu-Ru nanotubes

red curve in Fig. 2.14c). All of these evidence showed that too large a lattice mismatch would effectively hinder the nucleation of hcp Ru on the $PdCu_3(001)$ substrate, causing a polycrystal Ru overlayer (Fig. 2.14d). In principle, the mechanism for the instability of an interface induced by the tensile surface-stress was very similar to the Asaro–Tiller–Grinfeld (ATG) instability [32, 33]. In addition, the analysis based on DFT calculations also indicates that for other Pd–Cu alloys, the advantage of

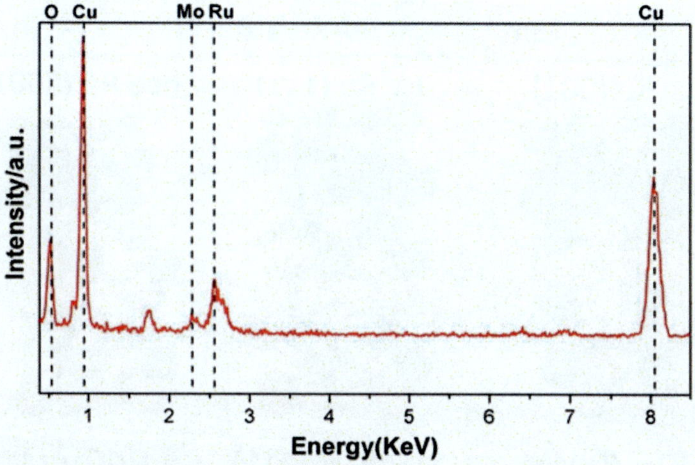

Fig. 2.20 EDS spectra of Cu-Ru nanotube

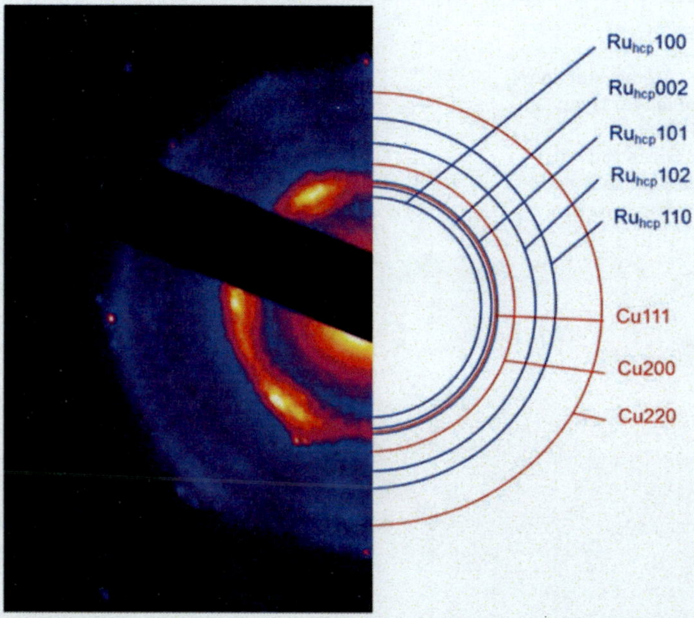

Fig. 2.21 Electron diffraction of Cu-Ru nanotube

the (smaller) lattice mismatch (and/or stronger binding energy) allows the epitaxial growth of inherently stable hcp Ru overlayers rather than inherently unstable fcc Ru ones (Figs. 2.14b and 2.23). All of these features are consistent with the experimental observations.

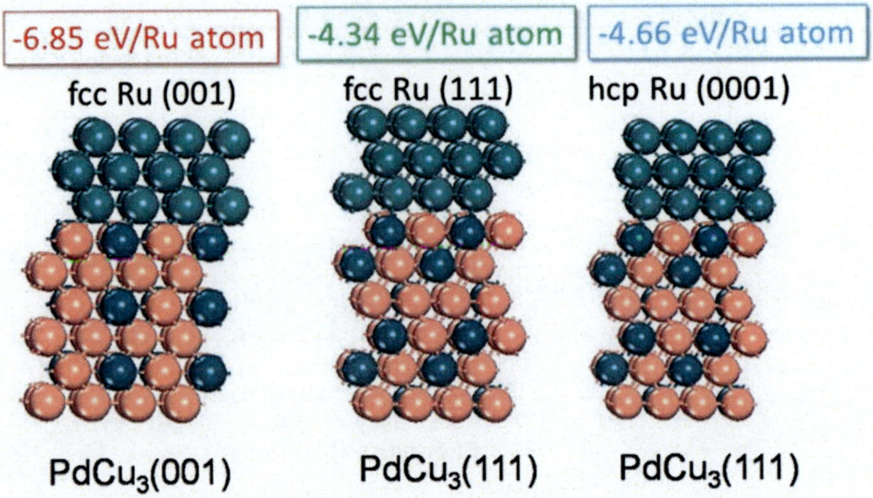

Fig. 2.22 Atomic structures of hcp and fcc Ru layers on the $PdCu_3$ (001) and (111) surfaces. Green, blue and red balls correspond to Ru, Pd and Cu atoms, respectively

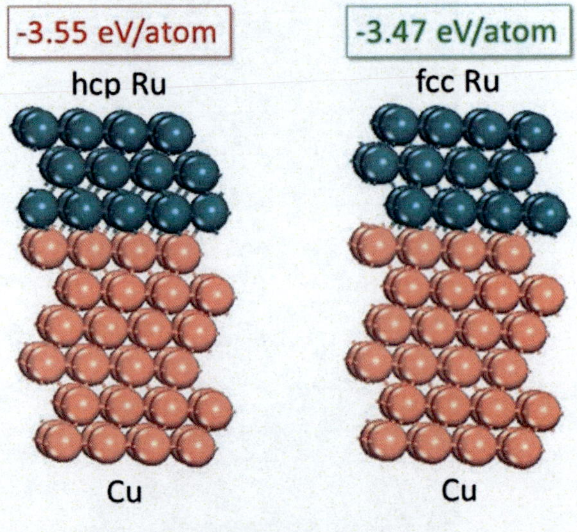

Fig. 2.23 Atomic structures of hcp and fcc Ru layers on the (111) face of Cu substrate. Green and red balls correspond to Ru and Cu atoms, respectively

To further explore the structure–activity relationship between catalytic behavior and crystallographic phase, hcp dominated Ru catalysts prepared from Cu NPs (defined as hcp Ru) were compared with fcc-dominated Ru. We performed the hydrogenation of 4-nitrochlorobenzene and styrene as probe reactions to investigate these two samples. In the hydrogenation of 4-nitrochlorobenzene, the fcc Ru catalyst gave more activity with over 99% conversion of 4-nitrochlorobenzene in 60 min. In contrast, only 61% conversion was achieved by the hcp-dominated Ru

(Fig. 2.14e). However, the opposite trend, that hcp Ru exhibit higher activity than fcc Ru was observed in the hydrogenation of styrene (Fig. 2.14f). The conversion of styrene was over 98% catalyzed by hcp Ru NPs compared with 53% conversion with fcc Ru NPs catalyst. This catalytic difference may be a result of the different adsorption behavior of the substrates on fcc and hcp Ru NPs. Importantly, in the semi-hydrogenation of diphenylacetylene the over-reduction product can be largely depressed by both the fcc and hcp Ru, which may result from poisoning effects by the generation of an Ru substrate complex on the surface. Moreover, the stereo selectivities of cis-stilbene can be maintained both by the hcp and fcc Ru catalysts (Figs. 2.24 and 2.25).

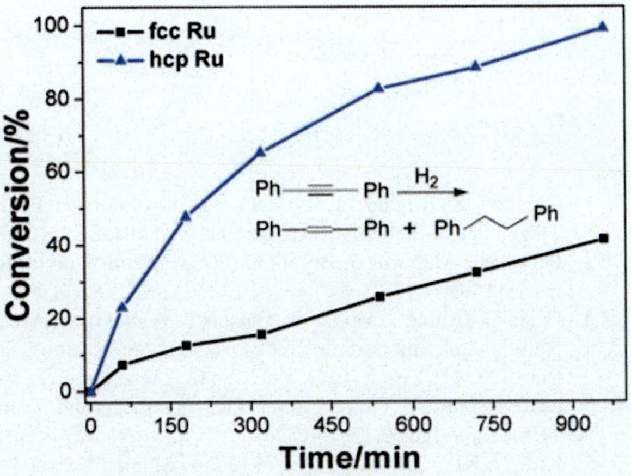

Fig. 2.24 Semi-hydrogenation of diphenylacetylene catalyzed by hcp Ru and fcc Ru

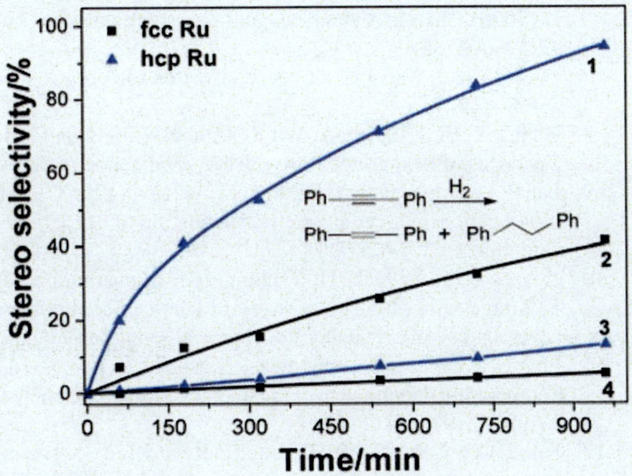

Fig. 2.25 Stereo selectivity of semi-hydrogenation of diphenylacetylene. Curve 1 and 2 are on behalf of cis-styrene and curve 3 and 4 are on behalf of trans-styrene

2.4 Conclusions

To conclude, it has been demonstrated that the lattice parameter plays a crucial role in controlling the crystallographic structure of Ru on the surface of Pd–Cu alloy. By optimizing the lattice parameter of the Pd–Cu alloy substrate, we are able to induce the unconventional fcc Ru overlayer by epitaxial growth, accompanied by a transformation from a trimetallic core–shell to yolk–shell structure. The trigger of epitaxial growth can be modulated by tuning the lattice mismatch between the alloy substrate and the overlayer, which enables the selective preparation of fcc or hcp Ru phases. Our findings provide a novel strategy to design multifunctional nanomaterials, simultaneously realizing composition, shape, and crystallographic structure control.

References

1. Nørskov JK, Bligaard T, Hvolbæk B, Abild-Pedersen F, Chorkendorff I, Christensen CH (2008) The nature of the active site in heterogeneous metal catalysis. Chem Soc Rev 37:2163–2171
2. Pelton M, Aizpurua J, Bryant G (2008) Metal-nanoparticle plasmonics. Laser Photonics Rev 2:136–159
3. Doria G, Conde J, Veigas B, Giestas L, Almeida C, Assunção M, Rosa J, Baptista PV (2012) Noble metal nanoparticles for biosensing applications. Sensors 12:1657–1687
4. Ling T, Xie L, Zhu J, Yu H, Ye H, Yu R, Cheng Z, Liu L, Liu L, Yang G (2009) Icosahedral face-centered cubic Fe nanoparticles: facile synthesis and characterization with aberration-corrected TEM. Nano Lett 9:1572–1576
5. Kim C, Kim C, Lee K, Lee H (2014) Shaped Ni nanoparticles with an unconventional hcp crystalline structure. Chem Commun 50:6353–6356
6. Tzitzios V, Basina G, Gjoka M, Alexandrakis V, Georgakilas V, Niarchos D, Boukos N, Petridis D (2006) Chemical synthesis and characterization of hcp Ni nanoparticles. Nanotechnology 17:3750–3755
7. Liu X, Luo J, Zhu J (2006) Size effect on the crystal structure of silver nanowires. Nano Lett 6:408–412
8. Huang X, Li S, Huang Y, Wu S, Zhou X, Li S, Gan C, Boey LF, Mirkin CA, Zhang H (2011) Synthesis of hexagonal close-packed gold nanostructures. Nat Commun 2:292–297
9. Fan Z, Huang X, Han Y, Bosman M, Wang Q, Zhu Y, Liu Q, Li B, Zeng Z, Wu J (2015) Surface modification-induced phase transformation of hexagonal close-packed gold square sheets. Nat Commun 6:6571–6579
10. Kusada K, Kobayashi H, Yamamoto T, Matsumura S, Sumi N, Sato K, Nagaoka K, Kubota Y, Kitagawa H (2013) Discovery of face-centered-cubic ruthenium nanoparticles: facile size-controlled synthesis using the chemical reduction method. J Am Chem Soc 135:5493–5496
11. Pust L, Wenger LE, Lukaszew RA, Sheng Y, Litvinov D, Wang Y, Uher C, Clarke R (1999) Temperature dependence of the magnetization reversal in Co(fcc)-BN-Co (poly hcp) structures. J Appl Phys 85:5765–5767
12. Kim J, Lee Y, Sun S (2010) Structurally ordered FePt nanoparticles and their enhanced catalysis for oxygen reduction reaction. J Am Chem Soc 132:4996–4997
13. Liu J-X, Su H-Y, Sun D-P, Zhang B-Y, Li W-X (2013) Crystallographic dependence of CO activation on cobalt catalysts: HCP versus FCC. J Am Chem Soc 135:16284–16287
14. Wu J, Li P, Pan Y-TF, Warren S, Yin X, Yang H (2012) Surface lattice-engineered bimetallic nanoparticles and their catalytic properties. Chem Soc Rev 41:8066–8074

15. Li Y, Cheng H, Yao T, Sun Z, Yan W, Jiang Y, Xie Y, Sun Y, Huang Y, Liu S (2012) Hexane-driven icosahedral to cuboctahedral structure transformation of gold nanoclusters. J Am Chem Soc 134:17997–18003
16. Manna L, Scher EC, Li L-S, Alivisatos AP (2002) Epitaxial growth and photo-chemical annealing of graded CdS/ZnS shells on colloidal CdSe nanorods. J Am Chem Soc 124:7136–7145
17. Liu J, Qiao SZ, Chen JS, Lou XWD, Xing X, Lu GQM (2011) Yolk/shell nanoparticles: new platforms for nanoreactors, drug delivery and lithium-ion batteries. Chem Commun 47:12578–12591
18. He D, Han Y, Fennell J, Horswell S, Li Z (2012) Growth and stability of Pt on Au nanorods. Appl Phys Lett 101:113102–113105
19. Gu J, Guo Y, Jiang Y-Y, Zhu W, Xu Y-S, Zhao Z-Q, Liu J-X, Li W-X, Jin C-H, Yan C-H (2015) Robust phase control through hetero-seeded epitaxial growth for face-centered cubic Pt@Ru nanotetrahedrons with superior hydrogen electro-oxidation activity. J Phys Chem C 119:17697–17706
20. Zhang J, Tang Y, Lee K, Ouyang M (2010) Nonepitaxial growth of hybrid core-shell nanostructures with large lattice mismatches. Science 327:1634–1638
21. Somorjai GA, Li Y (2010) Introduction to surface chemistry and catalysis. Wiley
22. Joo SH, Park JY, Renzas JR, Butcher DR, Huang W, Somorjai GA (2010) Size effect of ruthenium nanoparticles in catalytic carbon monoxide oxidation. Nano Lett 10:2709–2713
23. Dagle RA, Wang Y, Xia G-G, Strohm JJ, Holladay J, Palo DR (2007) Selective CO methanation catalysts for fuel processing applications. Appl Catal A 326:213–218
24. Macdonald JE, Sadan MB, Houben L, Popov I, Banin U (2010) Hybrid nanoscale inorganic cages. Nat Mater 9:810–815
25. Highlights P (2013) Platinum alloys: a selective review of the available literature. Platinum Metals Rev 57:202–213
26. Wu Y, Wang D, Niu Z, Chen P, Zhou G, Li Y (2012) A strategy for designing a concave Pt-Ni alloy through controllable chemical etching. Angew Chem Int Ed 51:12524–12528
27. Yin Y, Rioux RM, Erdonmez CK, Hughes S, Somorjai GA, Alivisatos AP (2004) Formation of hollow nanocrystals through the nanoscale Kirkendall effect. Science 304:711–714
28. Denton AR, Ashcroft NW (1991) Vegard's law. Phys Rev A 43:3161–3164
29. Danescu A (2001) The Asaro-Tiller-Grinfeld instability revisited. Int J Solids Struct 38:4671–4684
30. Stangl J, Holý V, Bauer G (2004) Structural properties of self-organized semi-conductor nanostructures. Rev Mod Phys 76:725–783
31. Trampert A (2002) Heteroepitaxy of dissimilar materials: effect of interface structure on strain and defect formation. Physica E: Low-Dimens Syst Nano-struct 13:1119–1125
32. Jesson D, Chen K, Pennycook S, Thundat T, Warmack R (1996) Morphological evolution of strained films by cooperative nucleation. Phys Rev Lett 77:1330–1333
33. Grinfel'd M (1991) Thermodynamic methods in the theory of heterogeneus systems. Longman Scientific and Technical

Chapter 3
Engineering the Electronic Structure of Single Atom Ru Sites via Compressive Strain Boosts Acidic Water Oxidation Electrocatalysis

3.1 Introduction

By developing efficient and durable electrolysers for water split ting, the intermittent electrical energy generated from renew able wind and solar energy can largely be converted into fuels [1]. Water electrolysis using a polymer electrolyte membrane (PEM), based on proton transfer [2], has been demonstrated to effectively mitigate the drawbacks of an alkaline environment, including the crossover of product gases, limited current density and low operating pressure [3–5]. Unfortunately, the sluggish catalysed oxygen evolution reaction (OER) and intense degradation of the catalysts [6, 7] in low pH and the strong oxidative environment imped the widespread adoption of practical electrolysers [8]. Compared with Ir-based systems, which have better dissolution resistance [9, 10], Ru has more abundant reserves and has been evaluated to be a more active OER catalyst due to its lower overpotential [11, 12]. The high activity of Ru-based catalysts for the OER in acidic conditions has been linked to its flexible redox state [13] due to the wide modulation spacing in response to the change in valence state induced by the frequent adsorption/desorption of oxygenated species [14]. This was corroborated by the simultaneous emergence of high-valence Ru species and corresponding high OER activity [15, 16]. The moderate oxygen bonding on Ru and low oxygen bulk diffusivity also contribute to its OER activity [17].

Typically, transient dissolution arising from structural disturbance related to frequent changes in the redox state of metals is responsible for their degradation [11]. This path is linked to the adsorbate evolution mechanism (AEM) of molecular oxygen generation. However, the accompanying over-oxidation of Ru to Ru–O_x moieties during the generation of (H)O ligands as a result of the demanding oxidative environment under electrode potentials is believed to trigger another dissolution route, where active oxygen-coordinated Ru moieties peel off from defects [18]. This represents a primary bottleneck hindering the application of Ru-based catalysts [15]. As shown in Fig. 3.1, these defects are mainly oxygen vacancies arising from the oxidative release of lattice oxygen to form molecular oxygen (the so-called lattice

Y. Yao, *Controllable Synthesis and Atomic Scale Regulation of Noble Metal Catalysts*, Springer Theses, https://doi.org/10.1007/978-981-19-0205-5_3

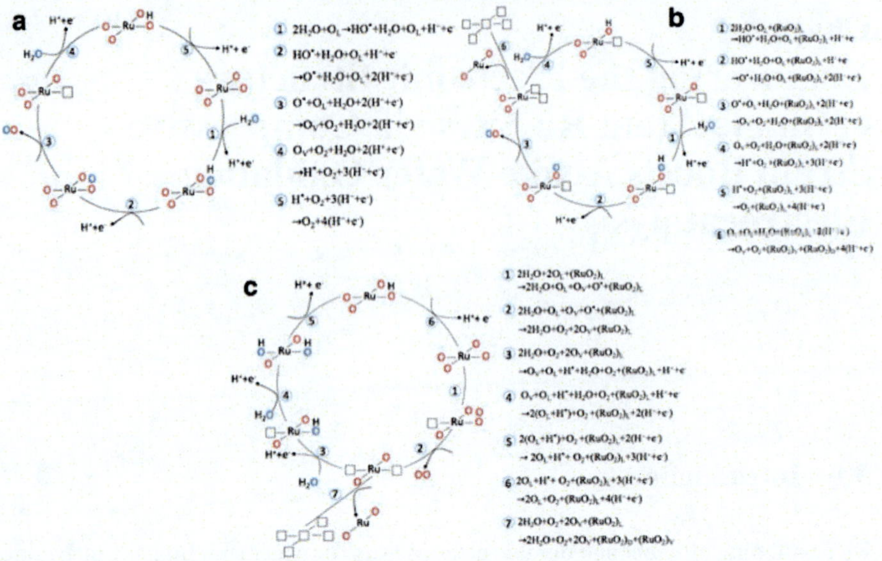

Fig. 3.1 **a** The formation of oxygen vacancy due to one O atom in generated O_2 molecule coming from lattice. **b** On the $RuO_2(110)$ with oxygen vacancy, further lose of lattice oxygen involved in OER process leads to the dissolution of Ru as RuO_2 molecule. **c** The lose of Ru due to the direct evolution of two oxygen atoms through the lattice distortion of lattice oxygen

oxygen evolution reaction, LOER) [19, 20]. In this dominant dissolution mechanism, as a result of the LOER process, the easy insertion/removal of oxygen in and out of its lattice is significant [21, 22]. Indeed, this rationalizes why thermally synthesized RuO_2, with its better crystallinity and fewer defects [23, 24], has a lower dissolution rate than electrochemically prepared RuO_2 [16, 22]. The rational design of a highly efficient and dissolution resistant Ru-based OER catalyst requires precise control over the electronic structure of Ru, preventing the dissolution of active Ru species in acid electrolytes and further lowering the overpotential for final industrial applications.

A reasonable means with which to lower the contribution of the LOER to the dissolution of Ru may involve switching the dominant O_2 generation mechanism from the LOER to the AEM [24]. This can be implemented by impeding the fast bulk diffusion rates and surface exchange kinetics of atomic oxygen by changing the coordination environment of the Ru. Metallic Pt has a significantly lower dissolution rate than Ru due to the relatively weaker oxygen bonding, resulting in a lower contribution of the LOER [11]. Embedding single atomic Ru into a Pt-rich coordination environment with higher corrosion potential may help improve the dissolution resistance of Ru by suppressing any local over-oxidation of Ru surface atoms. In addition, the atomic isolation of Ru without any formation of Ru–O–Ru oxygen bridging bonds maximizes the effect in modulating the coordination of Ru. Previous studies have shown that electrophilic oxygen ligand holes are induced by cationic lattice defects or low-valence metal ion dopants thanks to there being more covalent bonding, which prevents electrons from being transferred from the Ru/Ir centres to their oxygen ligands [25]. This mechanism will also help keep the resultant surface-exposed Ru sites in lower redox states, given the relatively lower valence state of Pt than Ru [15, 26]. A huge improvement in the isolated Ru's OER activity with the AEM mechanism, without compromising its stability, is further required, due to the failure of the traditional PtRu alloy in the OER activity [27]. The thin Pt shell environment, subject to compressive surface strain induced by the lattice mismatch, makes this possible [28, 29]. The magnitude of this strain can be controlled by the lattice parameter of the underlying substrate [30], through electrochemical leaching or acid etching of the Pt-based alloy. We consider it feasible that single Ru atoms anchored onto strained Pt overlayers may offer a method to efficiently engineer the electronic properties and OER reactivity with respect to redox chemistry.

We have synthesized a series of $PtCu_x/Pt_{skin}$ core–shell structures with atomically dispersed Ru_1 and have unravelled their mechanism of formation, oxidation resistance and the origin of the enhanced OER catalysis by exhaustively examining their structures, coordination environments and oxidation states. Through sequential acid etching and electrochemical leaching, the structure of $PtCu_x$ alloys can be varied (to give $PtCu_3$, PtCu and Pt_3Cu), which modulates effectively the OER activity catalysed by the Ru_1. The best catalyst, Ru_1–Pt_3Cu, delivers 220 mV overpotential to achieve a current density of 10 mA/cm^2 for acidic OER, with ten times longer lifetime over commercial RuO_2. We found that there is a volcano-type relation between the OER activity and the lattice constant. We argue that the compressive strain of the Pt_{skin} shell effectively engineers the electronic structure and redox behavior of single atomic Ru anchored at the corner or step sites of the Pt-rich shell, with optimized binding of oxygen intermediates and better resistance to over-oxidation and dissolution.

3.2 Experimental Section

3.2.1 Materials

Name	Formula	Purity	Manufacture
Ruthenium chloride	$RuCl_3 \cdot xH_2O$	A.R.	Sigma-Aldrich
Platinum acetylacetonate	$Pt(acac)_2$	A.R.	Alfa Aesar
Copper chloride	$CuCl_2 \cdot 2H_2O$	A.R.	Sinopharm
Ethanol	CH_3OH	A.R.	Sinopharm
Nitric acid	HNO_3	A.R.	Sinopharm
Hexane	C_6H_{14}	A.R.	Sinopharm
Oleylamine	OAm	A.R.	Sigma-Aldrich
Octadecene	ODE	A.R.	Sigma-Aldrich

(continued)

(continued)

Name	Formula	Purity	Manufacture
5% Nafion	Nafion	A.R.	Sigma-Aldrich
Argon	N_2	99.99%	Nanjing Special Gas Corp

The reagents were directly used without purification

3.2.2 Methods

Preparation of Cu nanowires. In a typical procedure for synthesizing Cu nanowire nanocrystals, $CuCl_2 \cdot 2H_2O$ (10 mg) and $RuCl_3 \cdot xH_2O$ were dissolved in 1 ml ethanol and 3 ml oleylamine. The mixture was sonicated for 30 min. The as-obtained suspension solution was slowly injected into a vial containing 3 ml 1-octadecylene, preheated at 120 °C for 10 min with vigorous stirring, then heated from 120 to 160 °C. The reaction last 6 h. After being cooled to room temperature, the products were precipitated by ethanol, separated via centrifugation and further purified by an ethanol–hexane mixture three times.

Preparation of Cu@Ru_1–$PtCu_3$. In a typical procedure, 5.65 mg of $Pt(acac)_2$ dissolved in 1 ml oleylamine was added to the as-prepared Cu nanowires and the temperature was changed from 160 to 200 °C followed by 12 h thermal treatment at 200 °C. The products were precipitated by ethanol, separated via centrifugation and further purified by an ethanol–hexane mixture several times.

Preparation of Ru_1–PtCu. In a typical procedure to obtain Ru_1–PtCu catalyst, Cu@Ru_1–$PtCu_3$ nanocrystals were dissolved in 10 ml 7.5 M nitric acid, followed by 2 h stirring at room temperature. Finally, the products were separated via centrifugation and further purified by ethanol.

Preparation of Ru_1–Pt_3Cu. These Ru_1–PtCu nanowires were loaded onto carbon black (XC-72) and annealed in an argon atmosphere at 400 °C for 12 h. The catalysts then underwent CV measurements in N_2-saturated 0.1 M $HClO_4$ solution (0.01–1.21 V vs. RHE) with a sweep rate of 100 mV/s.

3.2.3 Characterizations

Powder X-ray diffraction patterns of samples were recorded using a Rigaku Miniflex-600 with Cu Kα radiation (Cu Kα, $\lambda = 0.15406$ nm, 40 kV and 15 mA). The morphologies were characterized by TEM (Hitachi-7700, 100 kV). HAADF–STEM images and the corresponding electron energy-loss spectroscopy were collected on a JEOL JEM-ARM 200F TEM/STEM with a spherical aberration corrector working at 200 kV.

HAADF–STEM tomography was performed in a FEI Talos F200X operated at 200 kV. The electron micrographs were acquired automatically using a single-axis tilt tomography holder 2021 from −70° to 70°, with projections taken every 1° above 40°, and every 2° below 40°. During acquisition of the HAADF–STEM tilt series, the camera length was 98 mm. Stack alignment and reconstruction of the tile series were carried out offline using the software package Inspect 3D 4.1.2 (FEI). The aligned tilt series was reconstructed using the Simultaneous Iterative Reconstruction Technique (SIRT) with 100 iterations. The reconstructed structure was segmented manually and visualized in the software package Avizo Version 9.2.0.

Scanning electron microscopy (SEM) was carried out on a JSM-6700F SEM. Nitrogen sorption measurements were conducted using a Micromeritics ASAP 2020 system at 77 K. The XPS experiments were performed at the photoemission end-station at beamline BL 10B in the National Synchrotron Radiation Laboratory (NSRL) in Hefei, China. Briefly, the beamline was connected to a bending magnet and covered photon energies from 60 to 1000 eV with a resolving power (E/ΔE) better than 1000. In the current work, the structure of the samples was studied at photon energies of 1253.6 eV. Elemental analysis of Ru, Pt and Cu in the samples was conducted with an Optima 7300 DV for ICP-MS.

Ex situ XAFS measurements. XAFS spectra data (Ru K-edge, Pt L_3-edge and Cu K-edge) were collected at the 1W1B station in the Beijing Synchrotron Radiation Facility (BSRF, operated at 2.5 GeV with a maximum current of 250 mA) and the BL14W1 station in Shanghai Synchrotron Radiation Facility (SSRF, operated at 3.5 GeV with a maximum current of 250 mA). The data were collected at room temperature (25 °C) (Ru K-edge in fluorescence excitation mode using a seven-element Ge detector, Pt L_3-edge in fluorescence excitation mode using a Lytle detector and Cu K-edge in transmission mode using a N_2-filled ionization chamber). All samples were pelletized as disks of 13 mm diameter and 1 mm thickness using graphite powder as a binder.

In situ XAFS measurements. Electrochemical measurements were conducted on a computer-controlled electrochemical analyzer. Catalyst-modified carbon paper was used as the working electrode, Pt wire as the counter electrode and an Ag/AgCl (3 M KCl) electrode as the reference electrode. A home-made electrochemical cell was used for in-situ XAFS measurements. In-situ XAFS was used to obtain the change of valence state and coordination environment for elemental analysis of the catalyst during reactions. We prepared dilute catalyst solutions (2 mg/ml for 5 μl) and distributed them evenly on the center of carbon paper to form uniform films. The acquired EXAFS data were processed according to standard procedures using the ATHENA module implemented in the IFEFFIT software package. The EXAFS spectra were obtained by subtracting the post-edge background from the overall absorption and then normalizing with respect to the edge-jump step. Subsequently, the $\chi(k)$ data were Fourier-transformed to real (R) space using a Hanning window ($d_k = 1.0$ Å^{-1}) to separate the EXAFS contributions from different coordination shells. To obtain the quantitative structural parameters around the central atoms, least-squares curve parameter fitting was performed using the ARTEMIS module of the IFEFFIT software package.

In situ electrochemical attenuated total reflection Fourier transformed infrared spectroscopy (EC-ATR-FTIRS) measurements. A catalyst ink was first prepared by mixing 1 ml ethanol with 4 mg Ru_1–Pt_3Cu catalyst. The catalyst ink was then dropped via pipette onto a hemicylindrical silicon prism covered with several layers of graphene. A Pt foil and an Ag/AgCl electrode were used as counter- and reference electrodes, respectively. Millipore Milli-Q water (18.2 MΩ*cm) and ultrapure perchloric acid (70%, Suprapure, Sigma-Aldrich) were used to prepare the solution. The supporting electrolyte used in the measurement was 0.1 M $HClO_4$, which was constantly purged with N_2 (Nanjing Special Gas Corp) during the experiment. Before the experiments, the working electrode was cleaned by continuously scanning the electrode potential in the region 0.6–1.6 V for about 30 min. The electrode potential was held at 0.8 V in 0.1 M $HClO_4$ and a background spectrum (reflectance R_0) was recorded. The electrode potential was first altered from 0.8 to 1.6 V stepwise, then back again; in the meantime, infrared spectra were recorded with a time resolution of 5 s per spectrum at a spectral resolution of 4 cm^{-1}. All spectra were presented as absorbance, $A = -\log(R/R_0)$, where R is the reflectance of the sample spectrum. A Varian FTS 7000e infrared spectrometer with an HgCdTe detector cooled by liquid nitrogen was used.

3.2.4 Electrochemical Measurements

Electrochemical measurements for OER. Electrochemical tests were conducted on a CHI 760E electrochemical work station (Chenhua) with a three-electrode system in 0.1 M $HClO_4$ aqueous solution. The Ag/AgCl (3 M KCl) electrode was used as a reference electrode and a platinum wire was used as a counter electrode. A constant O_2 or N_2 flow was maintained in the headspace of the electrolyte during the entire experiment. The synthesized Ru_1–Pt_3Cu, Ru_1–PtCu and Cu@Ru_1–$PtCu_3$, PtRu nanoparticles (NPs) and PtCu NPs were dispersed onto carbon black (Cabot, Vulcan XC-72) with a metal loading of ~30%. The catalysts were subjected to a 400 °C annealing process for 12 h in argon to remove organic surfactants. The turbid catalyst ink was prepared by mixing 4 mg of the catalyst in 1 ml solution containing 960 μl ethanol and 40 μl 5% Nafion solution, followed by ultrasonication for 30 min. Similarly, the catalyst inks for commercial RuO_2 and IrO_2 samples were prepared by mixing 4 mg in 960 μl ethanol and 40 μl 5% Nafion solution as above. Next, a certain volume of the catalyst ink was carefully dropped onto a 5 mm diameter glassy carbon disk electrode or RRDE. The catalyst layer was allowed to dry at room temperature before electrochemical measurements. In 0.1 M $HClO_4$, the catalyst loading was 0.0163 mg_{Pt+Ru}/cm^2 and the loading of Ru was 0.00192 mg_{Ru}/cm^2 (based on ICP-MS). Electrochemical impedance spectroscopy measurements were conducted at

1.20 V (vs. Ag/AgCl) on a rotation electrode under 1600 rpm. The amplitude of the sinusoidal wave was 5 mV, and the frequency scan range was 100 kHz–0.1 Hz. Unless otherwise stated, all experiments were performed at ambient temperature (25 °C) and all potentials were referenced to the RHE. CV measurements were carried out in 0.1 M $HClO_4$ solutions under a flow of N_2 at a sweep rate of 50 mV/s. The specific ECSA was calculated by measuring the H_{upd} adsorption charge (Q_H) collected in the hydrogen adsorption/desorption region (0.01–0.31 V) after double-layer correction, according to the following relation:

$$ECSA = \frac{Q_H}{m * q_H}$$

where Q_H is the charge for H_{upd} ($H^+ + e^- \rightarrow H_{upd}$) adsorption, m is the metal loading and q_H (210 $\mu C/cm^2$) is the charge required for monolayer adsorption of hydrogen on Pt surfaces.

CO stripping measurement. For CO stripping, pure CO gas was purged through the catalyst surface in cells filled with 0.5 M H_2SO_4 electrolyte for 30 min while holding the working electrode at 0.1 V (vs. RHE). After transferring the electrodes to another cell filled with fresh 0.5 M H_2SO_4 electrolyte (without CO), CO stripping was performed in the potential range 0–1.2 V (vs. RHE) at a scan rate of 10 mV/s. The $ECSA_{CO}$ was also calculated using equation.

Rotating disk electrode voltammetry. The electrochemical experiments were conducted in O_2-saturated 0.1 M $HClO_4$ for the OER at room temperature at 1600 rpm. For RRDE tests, a computer-controlled CHI 760E electrochemical workstation was used, the disk electrode was scanned at a rate of 10 mV/s and the ring electrode potential was set to 1.48 V versus RHE. The hydrogen peroxide yield ($H_2O_2\%$) and the electron transfer number (n) were determined by the following equations:

$$H_2O_2\,(\%) = 100 \times \frac{2 * \frac{I_r}{N}}{I_d + \frac{I_r}{N}}$$

$$n = \times \frac{I_d}{I_d + \frac{I_r}{N}}$$

where I_d is the disk current, I_r is the ring current and $N = 0.4$ is the current collection efficiency of the Pt ring.

Stability measurement. The accelerated durability tests of the catalysts were performed in the O_2-saturated 0.1 M $HClO_4$ electrolyte at room temperature (25 °C) by applying potential cycling between 1.0 and 1.6 V versus Ag/AgCl at a sweep rate of 50 mV/s for 3000 cycles.

3.3 Results and Discussion

Uniform fivefold twinned Cu nanowires with an average diameter of 22 nm were prepared based on a previously established method (Supplementary Fig. 3.2) [31]. Ru-doped $PtCu_3$ nano-island chains were then generated epitaxially using electroless galvanic redox replacement in the presence of Pt and Ru ions, resulting in the growth of metallic islands consisting of dispersed Ru atoms embedded in a Cu-rich alloy (Fig. 3.3a). Island growth occurred largely from the five-fold twin boundaries of the pristine Cu nanowires. Epitaxial growth and the resultant nano-island formation of $PtCu_3$ from the Cu nanowires were corroborated by the identical atomic stacking mode (interplanar spacing and crystal facet orientation) of the island and the underlying substrate, as determined from aberration corrected high-angle annular dark-field scanning transmission electron microscopy (HAADF–STEM) imaging with sub-ångstrom resolution and fast Fourier transform (FFT) patterns (Fig. 3.4). Treatments at elevated temperatures from 433 to 473 K initiated an alloying process between Pt domains and the Cu substrate as well as a rearrangement of the individual Ru atoms as dopants. As a result, the formation of ordered alloy phases was observed. The observed preferential growth along the twin boundaries might result from a stronger interface interaction of the defect-rich stepped surfaces of twin boundaries [32] compared with the other exposed lateral facets of the Cu nanowire. The narrow width of the twin boundaries of the Cu nanowire appear to effectively prevent a lateral

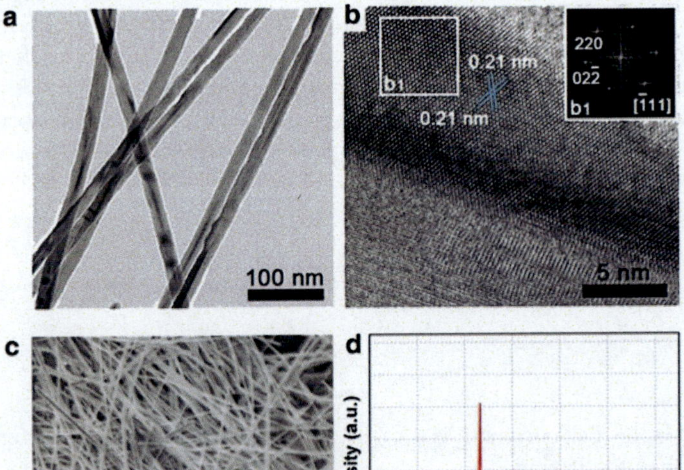

Fig. 3.2 Characterizations of Cu nanowires. **a** TEM image. **b** High magnification image and the corresponding FFT pattern. **c** SEM image. **d** XRD spectra

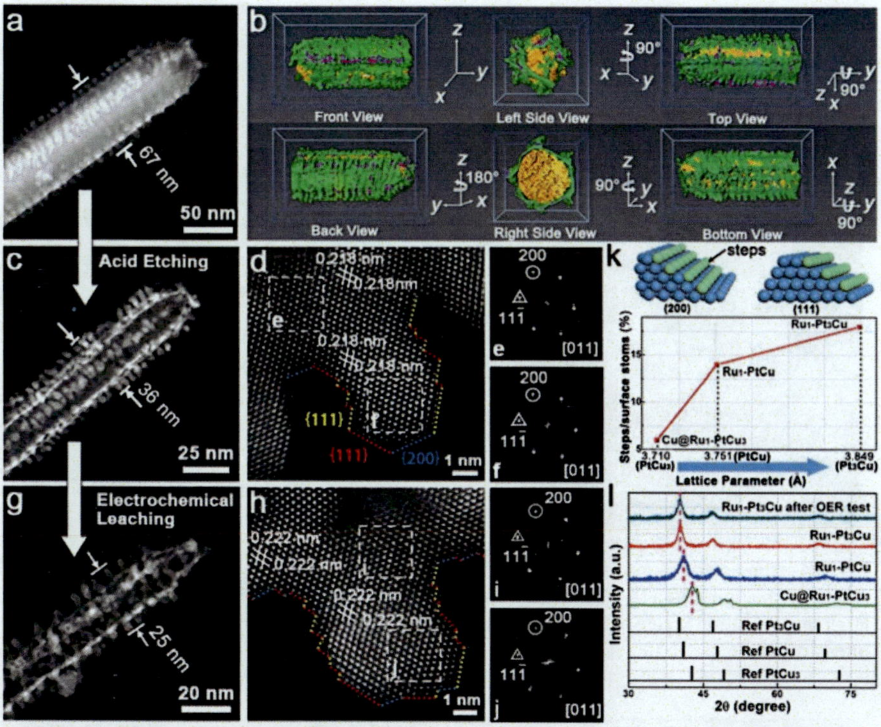

Fig. 3.3 Characterizations of $Cu@Ru_1–PtCu_3$, $Ru_1–PtCu$ and $Ru_1–Pt_3Cu$. HAADF–STEM image (**a**) and six projected images of a three-dimensional visualization of tomographic reconstruction images (**b**) of $Cu@Ru_1–PtCu_3$. HAADF–STEM image (**c**), high-magnification image (**d**) and corresponding FFT patterns (**e**, **f**) of $Ru_1–PtCu$. HAADF–STEM image (**g**), high-magnification image (**h**) and corresponding FFT patterns (**i**, **j**) of $Ru_1–Pt_3Cu$. In **d** and **h**, red and yellow dashed lines represent {111} facets and blue dashed lines represent the {200} facet. **k** Statistics of steps exposed on different facets of three representative samples. **l** XRD spectra of three representative processes

extension of the $PtCu_3$ islands to other facets, further facilitating the formation of one-dimensional (1D) chains of islands. 3D tomographic reconstructions (Fig. 3.3b) from the 2D HAADF–STEM images clearly demonstrate the preferential growth of Ru domains along the five-fold twin boundaries of the Cu nanowire and the distribution of periodic Pt–Cu islands. More detailed information about the individual Ru, Pt and Cu tomographic images and animated voxel recordings of the rotation of the tomogram are available in Fig. 3.5.

To remove buried non-functional Cu in the bulk of the nanowires as well as non-alloyed Ru, and to provide better accessibility to surface active sites, the Cu nanowires with protruding $PtCu_3$ nanoisland chains were treated in nitric acid. The acid treatment transformed the island chains into an acid-stable Ru-doped $PtCu/Pt_{skin}$ tubular core–shell structure with five-fold rotational symmetry. The core composition transformed thereby from $PtCu_3$ to PtCu stoichiometry (Fig. 3.3c), as suggested by the identical interplanar expansion from 0.214 nm (Fig. 3.4e) to 0.218 nm

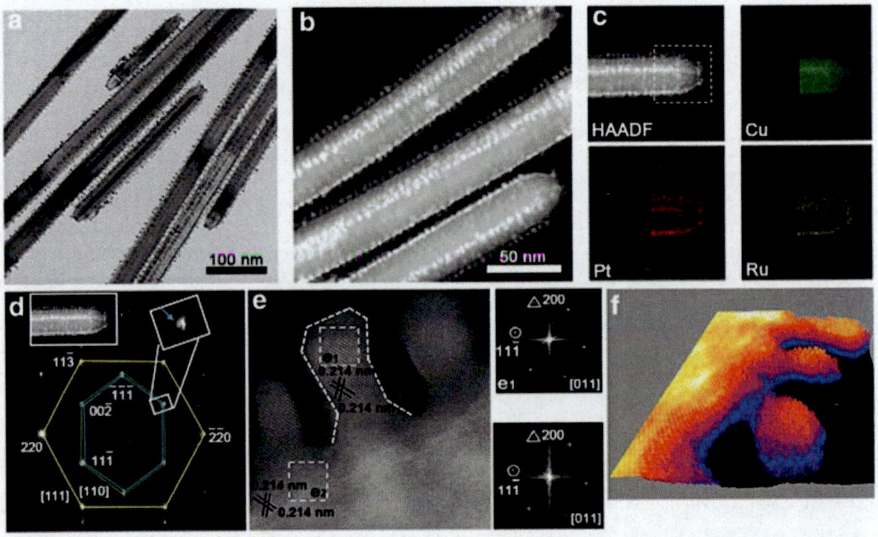

Fig. 3.4 Characterizations of Cu@Ru$_1$–PtCu$_3$. **a–c** TEM image, HAADF-STEM image and EDS elemental mapping. **d** HAADF-STEM image and corresponding electron diffraction pattern. **e** High magnification image and corresponding FFT patterns at different areas. **f** 3D HAADF-STEM intensity profile

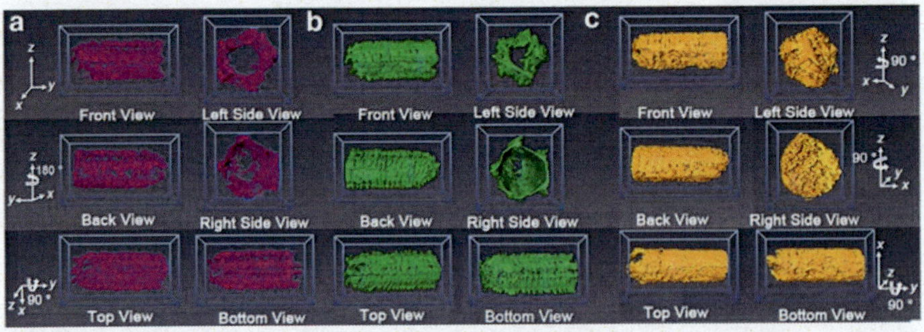

Fig. 3.5 Six projected images of three-dimensional visualization of tomographic reconstruction images of Cu@Ru$_1$–PtCu$_3$. The colors represent for different elemental, Ru (purple), Pt (green), and Cu (yellow), respectively

(Fig. 3.3d) in the nano-islands and underlying substrates. The dissolution of the Cu-rich nanowire core and the Ru–Cu distributed far from twin boundaries was supported further by contraction of the tubular structure diameter from 67 nm (Fig. 3.3a) to 36 nm (Fig. 3.3c). The Pt and Ru elemental distribution were retained after the acid etching, according to energy-dispersive X-ray spectroscopy (EDS) elemental mappings (Fig. 3.6), while the absolute Cu content obviously dropped sharply.

The elemental composition of the tubular dealloyed cores and their associated lattice spacing were further modulated by a subsequent electrochemical leaching

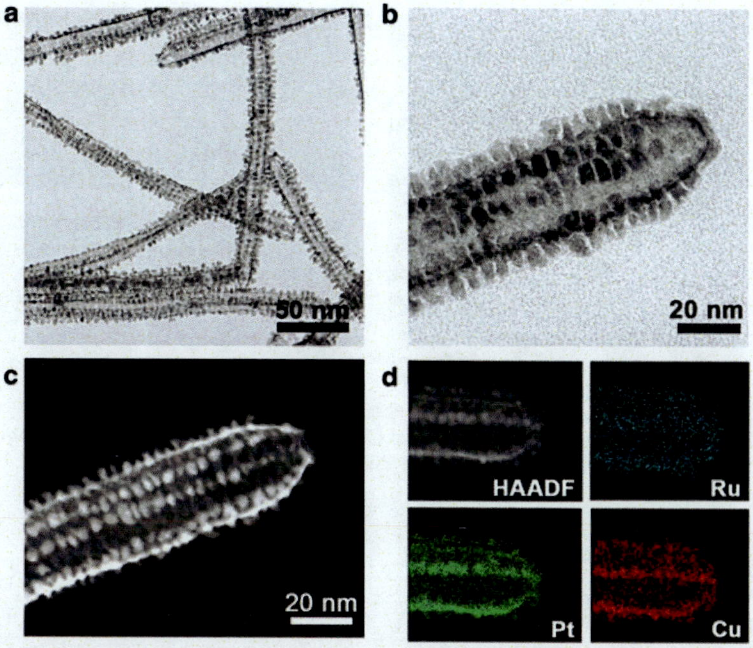

Fig. 3.6 Characterizations of Ru_1–PtCu. **a–c** TEM image, high magnification image, and HADDF-STEM image. **d** HAADF-STEM image and corresponding EDS elemental mapping. The selected area EDS elemental mapping clearly shows the islands-like distribution of Ru and Pt along the lateral twin boundaries of hollow tubular Ru_1–PtCu

process brought about by a cyclic voltammetry (CV) scan in N_2-saturated 0.1 M $HClO_4$ solution (+0.01 to +1.21 V vs. RHE) at a sweep rate of 50 mV/s. This voltammetric treatment removed additional Cu atoms and formed the final Ru-doped Pt_3Cu/Pt_{skin} core–shell nano–island chains (Fig. 3.7). In essence, the treatment served as an activation procedure for the catalyst. The principal morphology (Fig. 3.3g) and elemental distribution (Fig. 3.7) of the final nano-island chains were maintained during this CV activation process, although the tubular diameter contracted further from ~36 nm (Fig. 3.3c) to ~25 nm (Fig. 3.3g). The core composition (and corresponding interplanar spacing) changed from PtCu (Fig. 3.3d, 0.218 nm) to Pt_3Cu (Fig. 3.3h, 0.222 nm). The change in the ratio of Ru, Pt and Cu in three representative samples is shown in Table 3.1. An accurate count of the corner, step and kink defect sites exposed on the nano-island facets revealed that the acid etching and electrochemical leaching had increased the number of surface defects of the nano-island (Fig. 3.3k), which serve as anchor sites for dispersed individual Ru atoms (Ru_1). The compositional evolution of the alloy core of the tubular catalyst structure from $PtCu_3$ (green line, Fig. 3.3k) to PtCu (blue line) after acid treatment and further to Pt_3Cu (red line) after the electrochemical leaching process could be traced by the shift of the face-centred cubic (fcc) X-ray diffraction patterns (XRD) to smaller angles

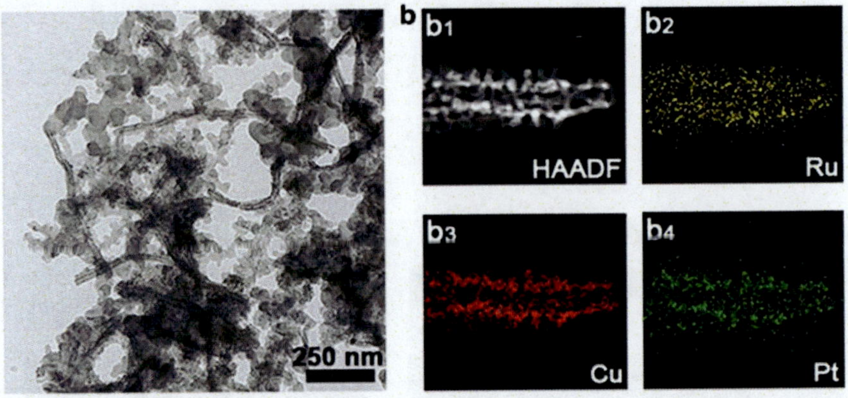

Fig. 3.7 Characterizations of Ru_1–PtCu after CV process. **a** TEM image. **b** EDS mapping of Ru_1–Pt_3Cu. The EDS mapping showed that the distributions of Ru, Pt, and Cu preserved after CV activation for Ru_1–Pt_3Cu

Table 3.1 The ratio of Ru, Pt and Cu in three representative samples by ICP-MS measurement

Sample	Ru (%)	Pt (%)	Cu (%)
Ru_1–$PtCu_3$	3.4	25.2	71.4
Ru_1–PtCu	5.6	46.3	48.1
Ru_1–Pt_3Cu	8.2	68.7	23.1

(Fig. 3.3l). In conclusion, a class of single-atom ternary alloys [33] were synthesized by acid etching and electrochemical leaching.

To further demonstrate the formation of the core–shell structure [30, 34] and atomic dispersion of Ru, atomically resolved elemental mapping [35] was performed to provide overall structural information, and typical results of the Pt_3Cu supported Ru are shown in edge structures, which cannot be found in Cu-resolved mapping (Fig. 3.8c). Superposition of the Pt and Cu mappings (Fig. 3.8e) showed an ordered atom arrangement and regular geometry profile, which are nearly identical to the corresponding HAADF image (Fig. 3.8a). The absence of Cu in the topmost layer (red dashed lines in Fig. 3.8e) provides clearer evidence of the formation of the Pt_{skin} shell. The formation of this Pt_{skin} shell explains the preservation of the XRD peak position (the dark cyan line in Fig. 3.3l) after 28 h OER test by preventing the bulk diffusion of inner Cu through the Pt_{skin} shell. Finally, negligible variations of the Cu valence state were observed by in situ K-edge X-ray absorption fine structure (XAFS) spectra with increasing working potentials. We believe that this was made possible by separation of the inner Cu species from oxygenated species by the Pt_{skin} shell (Fig. 3.8g). The Ru elemental distribution in Pt_3Cu shown in Fig. 3.8d indicates that the Ru is highly atomically dispersed and well separated in the Pt_3Cu/Pt_{skin} matrix, as marked by the arrows in Figs. 3.8f and 3.9.

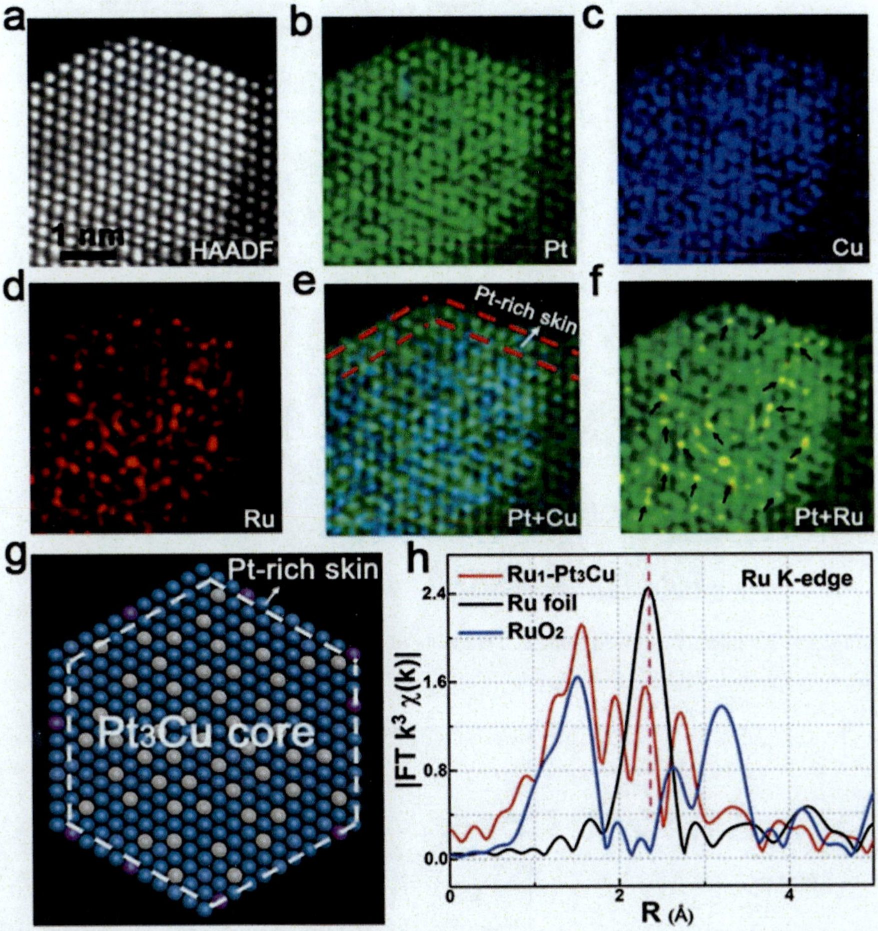

Fig. 3.8 Fine-structure characterizations of Ru_1–Pt_3Cu. **a–f** Atomically resolved elemental mapping of Ru_1–Pt_3Cu. **g** Schematic atom model of **a**. Blue, grey and purple represent Pt, Cu and Ru, respectively. **h** FT–EXAFS spectra of the Ru K edge

The way in which the atomic isolation and control of the redox state of Ru_1 influence the OER mechanism was further analyzed by clarifying its local coordination, using ex-situ and in-situ XAFS studies at the Pt and Ru edge. The Fourier transform (FT) of extended XAFS (FT–EXAFS) oscillations revealed that the Ru-doped Pt_3Cu exhibits one Fourier transform peak located at 1.58 Å attributed to the scattering of the Ru–O bond, while the peaks at 2.30 and 2.73 Å are attributed to the Ru–Pt and Ru–Cu bonds (Fig. 3.8h; for details see Supplementary Figs. 3.10, 3.11, 3.12, 3.13 and Tables 3.2, 3.3, 3.4). The absence of Ru–Ru bonds (at 2.38 Å) and related Ru–O–Ru coordination at a higher coordination shell excludes the presence of metallic Ru clusters and RuO_2 particles. This further supports the atomic isolation of Ru species

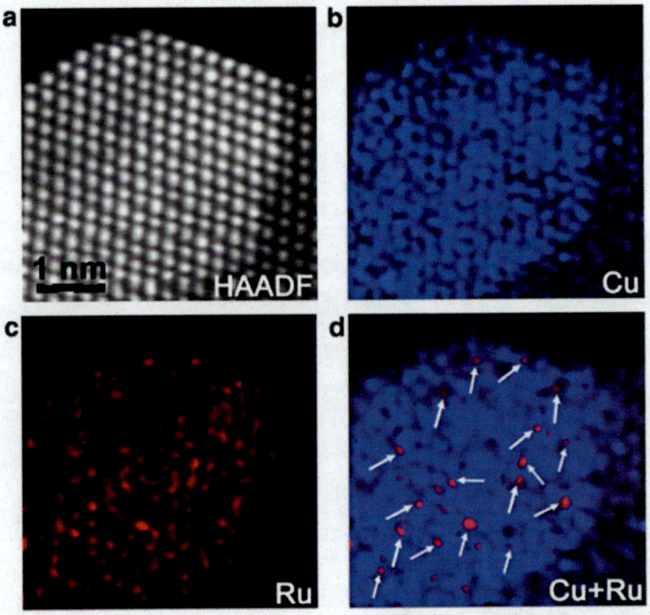

Fig. 3.9 Atomically resolved elemental mapping of Ru_1–Pt_3Cu electrocatalyst

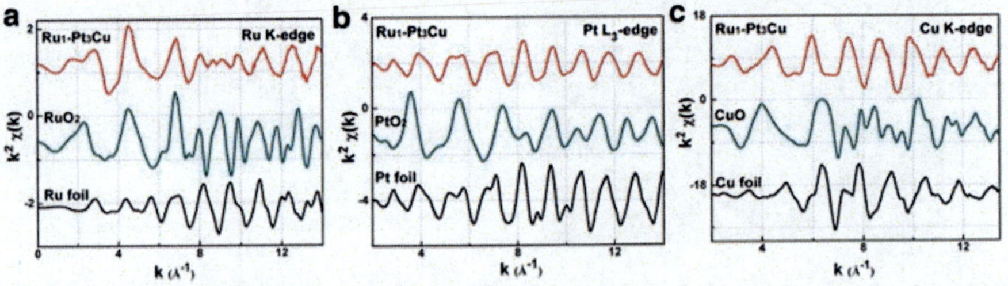

Fig. 3.10 EXAFS oscillations of Ru_1–Pt_3Cu. **a** Ru K-edge EXAFS oscillations of Ru_1–Pt_3Cu, RuO_2 bulk and Ru foil reference. **b** Pt L_3-edge EXAFS oscillations of Ru_1–Pt_3Cu, PtO_2 bulk and Pt foil reference. **c** Cu K-edge EXAFS oscillations of Ru_1–Pt_3Cu, CuO and Cu foil reference

dispersed inside the Pt_3Cu/Pt_{skin} core–shell particles (noted as Ru_1–Pt_3Cu henceforth). X-ray absorption near-edge structure (XANES) measurements showed that the corresponding Ru K-edge adsorption threshold of Ru_1–Pt_3Cu locates between the Ru foil and RuO_2 bulk (Fig. 3.14), indicating that surface-exposed Ru_1 sites were not over-oxidized.

The OER activities of Ru_1–$PtCu_3$, Ru_1–$PtCu$ and Ru_1–Pt_3Cu supported on active carbon (Fig. 3.15) were evaluated and compared to commercial RuO_2 and IrO_2 (Fig. 3.16) in a conventional three-electrode set-up, with O_2-saturated 0.1 M $HClO_4$ aqueous solution as the electrolyte. As revealed by the linear sweep voltammetry

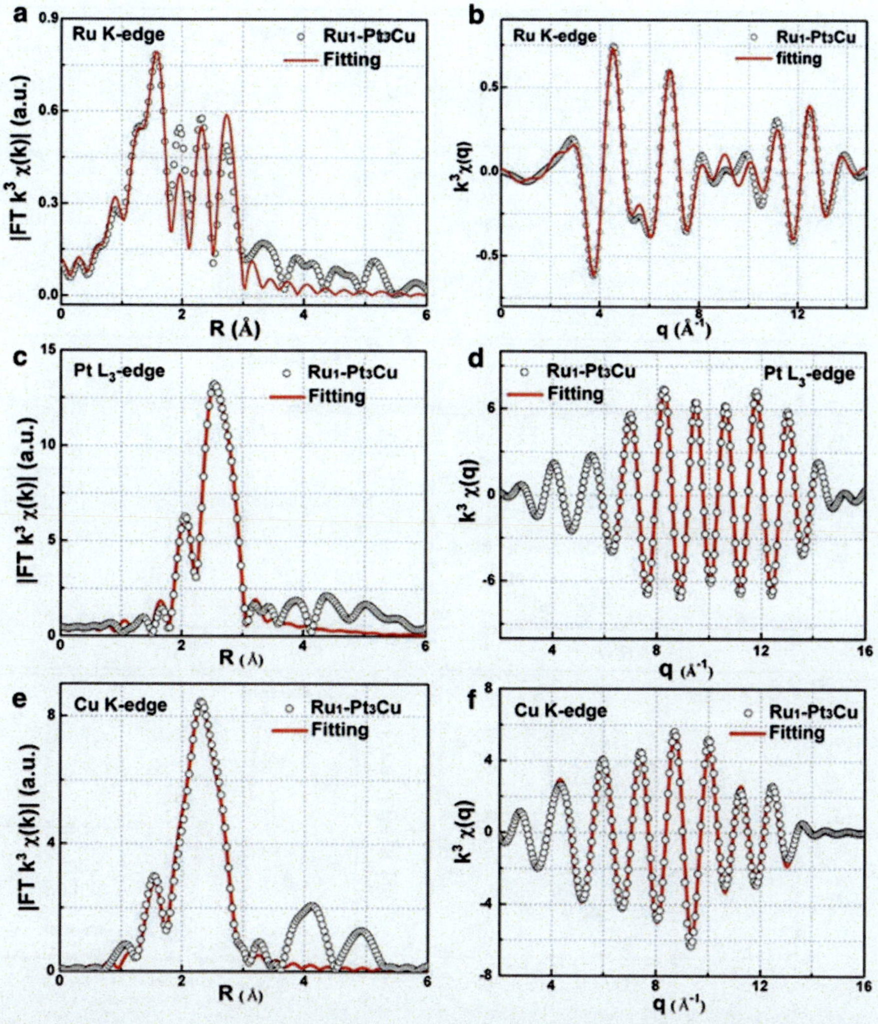

Fig. 3.11 R space and inverse FT-EXAFS fitting of Ru_1–Pt_3Cu catalyst. **a**, **b** R space and inverse FT-EXAFS fitting result of Ru K-edge (FT range: 2.0–11.0 $Å^{-1}$; fitting range: 0.5–3.5 Å), **c**, **d** Pt L_3-edge (FT range: 2.0–13.9 $Å^{-1}$; fitting range: 1.7–3.2 Å) and **e**, **f** Cu K-edge (FT range: 2.0–13.5 $Å^{-1}$; fitting range: 1.0–3.5 Å)

(LSV) presented in Figs. 3.17a and 3.18, Ru_1–Pt_3Cu has the catalytically most active polarization curve, with a sharp increase in anodic current density from an onset potential of 1.40 V (potential required to reach 0.1 mA/cm^2) referred to the reversible hydrogen electrode (RHE). The potential required to reach an OER current density of 10 mA/cm^2 is another key OER performance metric. A 280 mV overpotential was required for Ru_1–$PtCu_3$ to reach 10 mA/cm^2. With a decrease in Cu ratio, the corresponding overpotential decreased continuously to 250 mV for Ru_1–PtCu and

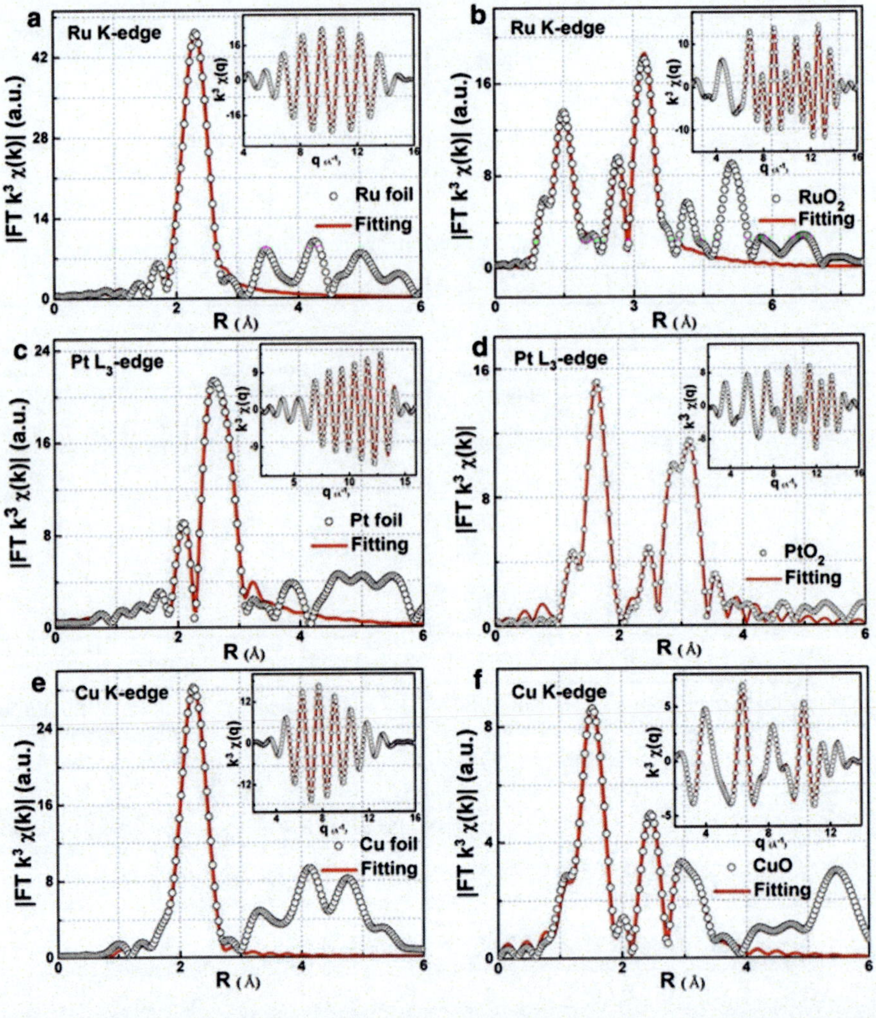

Fig. 3.12 R space and inverse FT-EXAFS fitting of control samples. **a**, **b** R space and inverse FT-EXAFS fitting result of Ru K-edge, Ru foil (FT range: 2.0–14.1 $Å^{-1}$; fitting range: 1.5–3.1 Å); RuO_2 (FT range: 2.0–14.1 $Å^{-1}$; fitting range: 0.8–4.0 Å). **c**, **d** Pt L_3-edge, Pt foil (FT range: 2.0–13.9 $Å^{-1}$; fitting range: 1.0–3.2 Å); PtO_2 (FT range: 2.0–14.4 $Å^{-1}$; fitting range: 1.0–3.5 Å) and **e**, **f** Cu K-edge, Cu foil (FT range: 2.0–13.4 $Å^{-1}$; fitting range: 1.4–3.0 Å); CuO (FT range: 2.0–12.8 $Å^{-1}$; fitting range: 0.6–3.9 Å)

220 mV for Ru_1–Pt_3Cu. However, for PtRu NPs without any Cu, the corresponding overpotential increased significantly to 340 mV (Figs. 3.17b and 3.19). We note that the OER activity of the PtCu alloy was rather poor, with an overpotential of 410 mV (Figs. 3.17a and 3.20). This suggests that the atomically dispersed Ru_1 centers embedded in the Pt–Cu alloys act as, or at least participate in, the active site for the OER. To assess the electrochemically active surface area (ECSA) we investigated

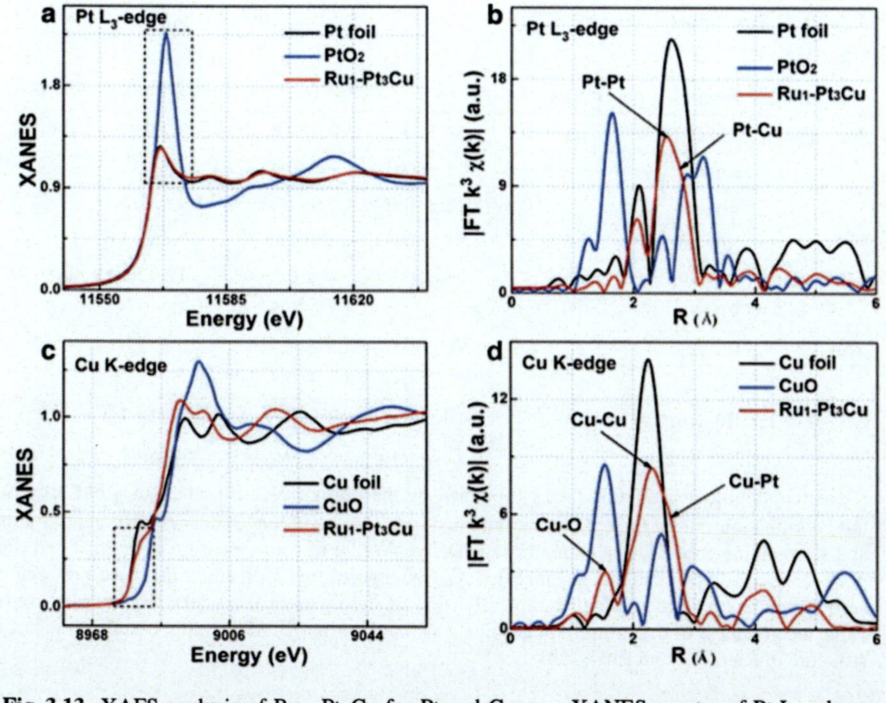

Fig. 3.13 XAFS analysis of Ru_1–Pt_3Cu for Pt and Cu. **a**, **c** XANES spectra of Pt L_3-edge and Cu K-edge. The oxidation state of Pt species and Cu species could be reflected by the white line intensity of Pt L_3-edge and the adsorption threshold of Cu K-edge, respectively [36, 37]. **b**, **d** FT-EXAFS spectra of Pt L_3-edge and Cu K-edge. We claimed that Cu species in Ru_1–Pt_3Cu were slightly oxidized after acid treatment, as a weak Cu–O scattering peak was monitored

hydrogen adsorption/desorption during CV as well as CO stripping (Figs. 3.21 and 3.22). The ECSAs calculated from the H_{upd} adsorption/desorption peak area (between 0.05 and 0.35 V) are consistent with the CO stripping results (Table 3.5).

The lattice constants of Ru_1–$PtCu_3$, Ru_1–PtCu, Ru_1–Pt_3Cu and PtRu NPs are plotted in Fig. 3.17b. It can be seen that the corresponding lattice constant increased with an increase in the ratio of Pt to Cu, as expected from Vegard's rule [41]. Interestingly, there is a volcano-type relation between overpotential (lower means more active, hence the volcano is inverse) and the lattice constant. The optimum lattice constant occurred for Ru_1–Pt_3Cu, which will be discussed in detail in the following. To the best of our knowledge, the OER overpotential of Ru_1–Pt_3Cu at the inverse-volcano peak outperformed the state-of-the-art RuO_2 (310 mV) and IrO_2 (380 mV) catalysts in acidic electrolyte (Table 3.6). Due to the atomic dispersion of the active Ru species, Ru_1–Pt_3Cu also shows a remarkable mass activity of 6615 A/g (normalized by Ru) and 779 A/g (normalized by Ru + Pt) at $\eta = 280$ mV, respectively (Fig. 3.21 and Table 3.7). Polarization curves at different temperatures for Ru_1–Pt_3Cu and RuO_2 were collected to assess the apparent kinetic barriers (Fig. 3.23). The derived Arrhenius plots were extended to the OER thermodynamic equilibrium

Table 3.2 Structural parameters of Ru_1–Pt_3Cu, Ru foil and RuO_2 bulk extracted from the EXAFS fitting ($^aS_0^2 = 0.85$)

Sample	Scattering pair	^{b}CN	^{c}R (Å)	$^d\sigma^2$ (10^{-3} Å^2)	$^e\Delta E_0$ (eV)	^{f}R factor
Ru_1–Pt_3Cu	Ru–O_1	1.1 ± 0.4	1.98 ± 0.02	3.9 ± 0.8	−9.2 ± 2.1	0.0054
	Ru–O_2	1.3 ± 0.5	2.05 ± 0.02	4.7 ± 0.9		
	Ru–Pt	3.2 ± 0.8	2.75 ± 0.02	4.5 ± 1.0	2.9 ± 1.2	
	Ru–Cu	2.6 ± 0.7	2.81 ± 0.02	4.7 ± 1.0	−5.2 ± 2.3	
Ru foil	Ru–Ru_1	6*	2.64 ± 0.02	3.1 ± 0.5	−12.6 ± 0.9	0.0010
	Ru–Ru_2	6*	2.72 ± 0.02	3.7 ± 0.5	1.8 ± 0.6	
RuO_2	Ru–O_1	2*	1.90 ± 0.02	3.3 ± 0.5	4.8 ± 1.0	0.0032
	Ru–O_2	4*	2.02 ± 0.02	3.7 ± 0.5		
	Ru–Ru_1	2*	3.10 ± 0.02	5.2 ± 0.6	−7.7 ± 1.2	
	Ru–Ru_2	8*	3.47 ± 0.02	5.5 ± 0.6	3.2 ± 0.5	

$^aS_0^2$ is the amplitude reduction factor (obtained by the fitting of Ru foil and RuO_2 bulk); ^{b}CN is the coordination number; ^{c}R is interatomic distance (the bond length between Ru central atoms and surrounding coordination atoms); $^d\sigma^2$ is Debye–Waller factor (a measure of thermal and static disorder in absorber–scatterer distances); $^e\Delta E_0$ is edge-energy shift (the difference between the zero kinetic energy value of the sample and that of the theoretical model); ^{f}R factor is used to value the goodness of the fitting. *This value was fixed during EXAFS fitting, based on the known structure of Ru metal and RuO_2 bulk

Table 3.3 Structural parameters of Ru_1–Pt_3Cu, Pt foil and PtO_2 bulk extracted from the EXAFS fitting ($^aS_0^2 = 0.90$)

Sample	Scattering pair	^{b}CN	^{c}R (Å)	$^d\sigma^2$ (10^{-3} Å^2)	$^e\Delta E_0$ (eV)	^{f}R factor
Ru_1–Pt_3Cu	Pt–Cu	3.1 ± 1.0	2.84 ± 0.2	5.2 ± 1.3	11.3 ± 3.0	0.0021
	Pt–Pt	8.6 ± 2.1	2.73 ± 0.2	4.3 ± 0.9	7.1 ± 2.5	
Pt foil	Pt–Pt	12*	2.76 ± 0.2	5.1 ± 0.2	7.1 ± 0.6	0.0013
PtO_2	Pt–O_1	2*	1.98 ± 0.2	3.1 ± 0.2	7.0 ± 1.0	0.0065
	Pt–O_2	4*	2.06 ± 0.2	4.2 ± 0.4		
	Pt–Pt	2*	3.11 ± 0.2	3.4 ± 0.3	14.2 ± 2.0	

$^aS_0^2$ is the amplitude reduction factor (obtained by the fitting of Pt foil and PtO_2 bulk); ^{b}CN is the coordination number; ^{c}R is interatomic distance (the bond length between Ru central atoms and surrounding coordination atoms); $^d\sigma^2$ is Debye–Waller factor (a measure of thermal and static disorder in absorber–scatterer distances); $^e\Delta E^0$ is edge-energy shift (the difference between the zero kinetic energy value of the sample and that of the theoretical model); ^{f}R factor is used to value the goodness of the fitting. *This value was fixed during EXAFS fitting, based on the known structure of Pt metal and PtO_2 bulk

Table 3.4 Structural parameters of Ru_1–Pt_3Cu Cu foil and CuO bulk extracted from the EXAFS fitting ($^aS_0^2 = 0.86$)

Sample	Scattering pair	^{b}CN	^{c}R (Å)	$^d\sigma^2$ (10^{-3} Å^2)	$^e\Delta E_0$ (eV)	^{f}R factor
Ru_1–Pt_3Cu	Cu–O	0.8 ± 0.3	1.97 ± 0.2	5.1 ± 1.2	3.9 ± 7.8	0.0038
	Cu–Cu	2.7 ± 0.8	2.61 ± 0.2	8.1 ± 4.3	−6.1 ± 4.8	
	Cu–Pt	3.5 ± 2.0	2.65 ± 0.2	8.2 ± 2.5	−6.1 ± 2.1	
Cu foil	Cu–Cu	12*	2.54 ± 0.2	8.2 ± 0.6	3.6 ± 0.6	0.0037
CuO	Cu–O_1	4*	1.95 ± 0.2	3.7 ± 0.9	3.5 ± 1.4	0.0013
	Cu–O_2	2*	3.02 ± 0.2	4.8 ± 0.5		
	Cu–Cu_1	4*	2.92 ± 0.2	6.3 ± 0.8	2.5 ± 0.4	
	Cu–Cu_2	4*	3.09 ± 0.2	4.6 ± 0.3	−2.5 ± 0.4	
	Cu–Cu_3	2*	3.13 ± 0.2	3.7 ± 0.3	−16.8 ± 3.5	
	Cu–Cu_4	2*	3.55 ± 0.2	9.6 ± 0.4	14.3 ± 11.4	

$^aS_0^2$ is the amplitude reduction factor (obtained by the fitting of Cu foil and CuO bulk); ^{b}CN is the coordination number; ^{c}R is interatomic distance (the bond length between Ru central atoms and surrounding coordination atoms); $^d\sigma^2$ is Debye–Waller factor (a measure of thermal and static disorder in absorber–scatterer distances); $^e\Delta E^0$ is edge-energy shift (the difference between the zero kinetic energy value of the sample and that of the theoretical model); ^{f}R factor is used to value the goodness of the fitting. *This value was fixed during EXAFS fitting, based on the known structure of Cu metal and CuO bulk

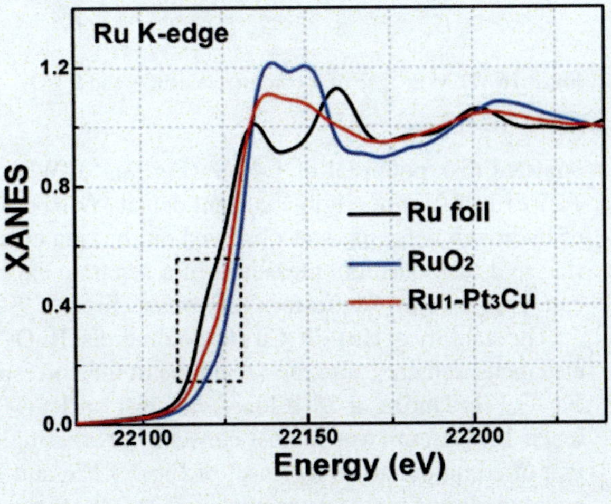

Fig. 3.14 XANES spectra of Ru_1–Pt_3Cu

potential, from which the apparent electrochemical activation energies (E_a) at η = 0 mV could be extracted [42]. As shown in Fig. 3.24, the extracted E_a for Ru_1–Pt_3Cu was as low as 30.8 kJ mol^{-1}, which is approximately half that of RuO_2 (61.9 kJ/mol).

Experiments on a rotating ring-disk electrode (RRDE) showed that the product catalyzed by Ru_1–Pt_3Cu were exclusively O_2. This was supported by the oxygen reduction reaction (ORR) polarization curve collected from the ring electrode at a

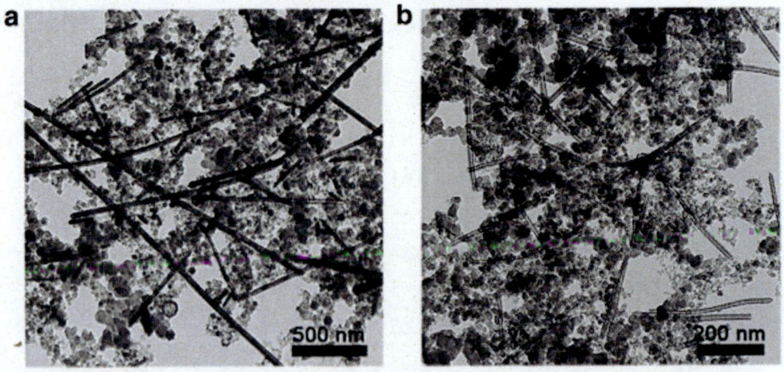

Fig. 3.15 TEM images of **a** Cu@Ru_1–$PtCu_3$ and **b** Ru_1–PtCu loaded on carbon black

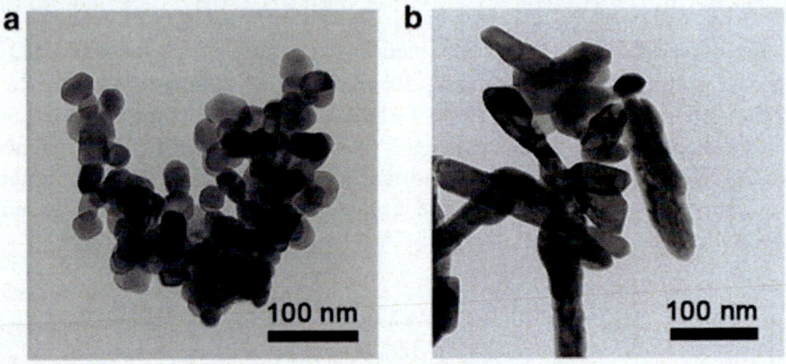

Fig. 3.16 TEM images of **a** commercial RuO_2 and **b** IrO_2

constant disk potential of 1.45 V (Fig. 3.25). When the disk potential varied from 1.27 to 1.48 V, a negligible current density (<10 μA/cm^2) assigned to the oxidation of hydrogen peroxide was observed on the ring electrode. According to the ratio of ring and disk currents measured in a separate experiment, the Faradaic efficiency catalyzed by Ru_1–Pt_3Cu for OER was as high as 99.8% (Fig. 3.26).

The stability of Ru_1–Pt_3Cu and commercial RuO_2 was tested by chronopotentiometry measurements, and the variations in the corresponding potentials are plotted in Fig. 3.17c. During a 28 h long-term test on Ru_1–Pt_3Cu, the potential required to reach 10 mA/cm^2 was almost constant (decreasing only by 2.1%), well surpassing that of commercial RuO_2 (inset of Figs. 3.17c and 3.27). We also used inductively coupled plasma mass spectrometry (ICP–MS) to quantify the mass loss of Pt, Cu and Ru during u–t measurements (Supplementary Fig. 3.28). The minor dissolved mass of Ru was much lower than the almost 100% dissolution of RuO_2 catalysts under acidic OER conditions after a test lasting approximately 2 h (inset of Fig. 3.17c). The stability was further demonstrated by the negligible activity decay in the polarization curve after 3000 cycles (Fig. 3.29). High-resolution transmission electron microscopy analysis on the recycled sample indicated that the island-like morphology

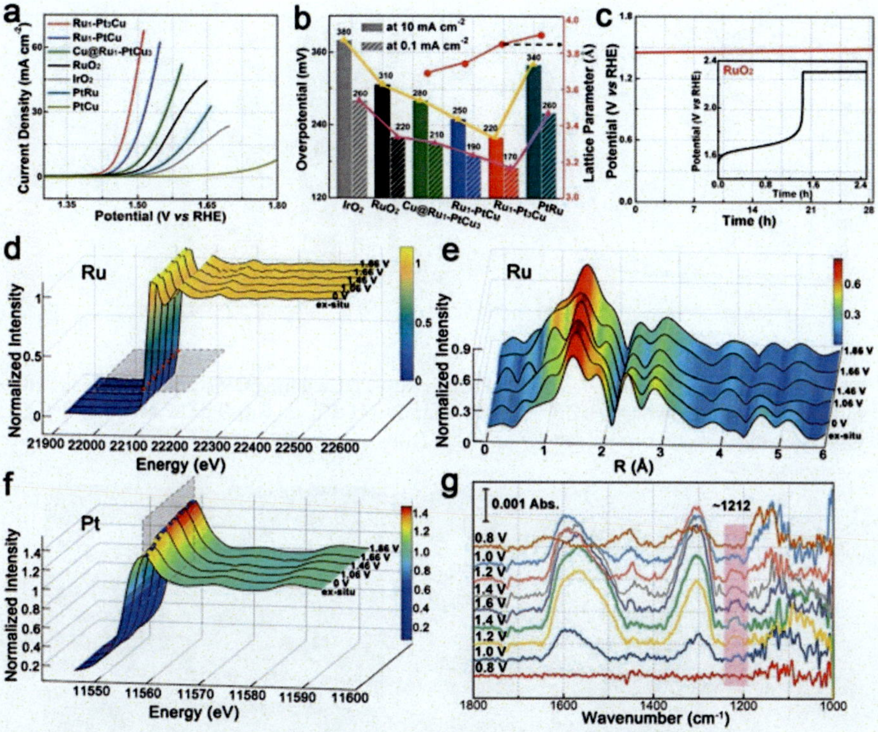

Fig. 3.17 Activity and stability investigations of Ru_1–Pt_3Cu during the OER process. **a** LSV curves of Ru_1–Pt_3Cu, Ru_1–PtCu, Cu@Ru_1–Pt_3Cu and reference samples in an O_2-saturated 0.1 M $HClO_4$ aqueous solution with 1600 rpm. **b** Overpotential to reach 0.1 and 10 mA/cm^2 for the catalysts (left axis) and lattice parameter dependence on the composition of Pt/Cu (right axis, red line). Error bars show the s.d. evaluated from five independent measurements. **c** Chronopotentiometric curve obtained with Ru_1–Pt_3Cu and commercial RuO_2 with constant current densities of 10 mA cm^{-2}. **d–f** In situ XAFS spectra change of the Ru K-edge (**d**, **e**) and Pt L_3-edge (**f**). **g** In situ ATR-IR spectra recorded during the multi-potential steps

and trimetallic interface were preserved to a large extent after the durability test (Fig. 3.30). The great stability of Ru_1–Pt_3Cu might arise from the unique periodic nano-islands providing abundant defect sites that anchor the atomically dispersed Ru_1 and prevent their migration and aggregation.

We carried out in-situ XAFS studies to investigate the potential-dependent oxidation of Ru_1 during the OER (Fig. 3.31). With the potential scanned from 0 to 1.86 V, the oxidation state of Ru_1 appeared stable, as shown by the essentially identical Ru_1 K-edge absorption threshold position (Figs. 3.17d and 3.32) and coordination (Fig. 3.17e). In contrast, an increase in the white line intensity of the Pt L_3-edge suggests increased d band vacancy for Pt atoms in the Pt–Cu domains (Figs. 3.17f, 3.32 and 3.33), probably because of the electron transferring from Pt to Ru_1. This is further corroborated by X-ray photoelectron spectroscopy (XPS) measurements, which revealed the change in electronic structure at the surface of the

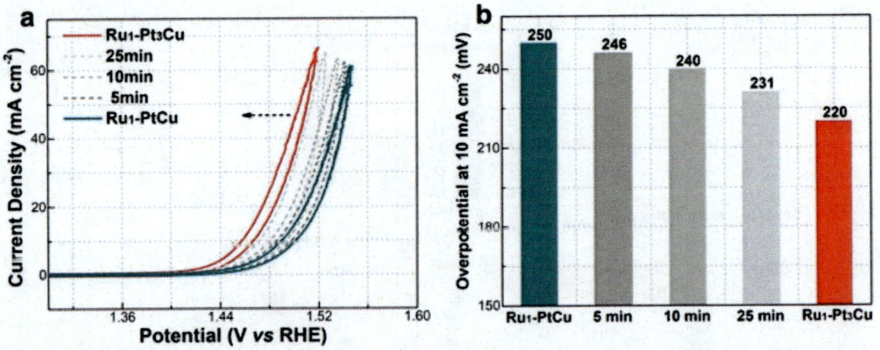

Fig. 3.18 The transformation from Ru_1–PtCu to Ru_1–Pt_3Cu during CV scanning. **a** Cyclic voltammograms (CVs) showing activity change from Ru_1–PtCu to Ru_1–Pt_3Cu as time changes. **b** Overpotential to reach current density of 10 mA/cm^2

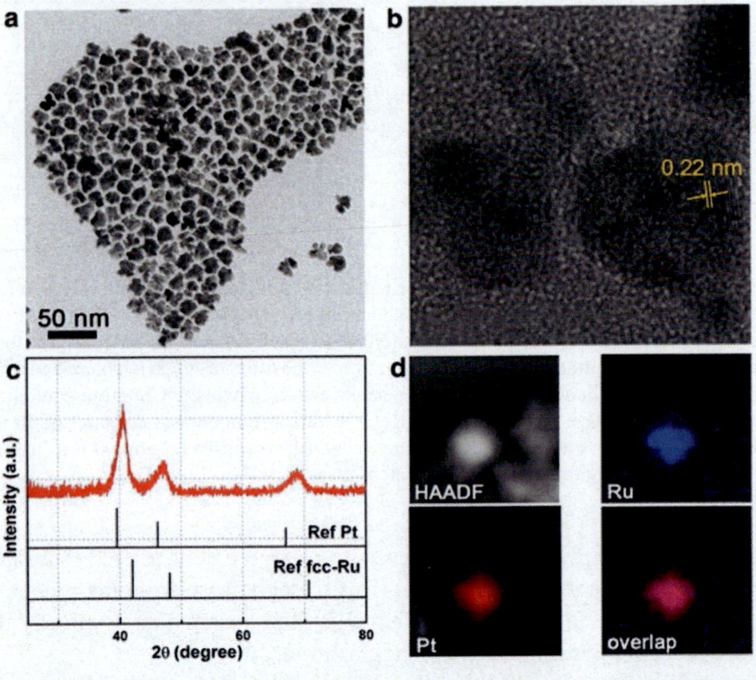

Fig. 3.19 Characterizations of PtRu NPs. **a**, **b** TEM and HRTEM images of PtRu. The TEM image of Pt–Ru alloy and reveals its irregular shape and the HRTEM image in confirms its lattice spacing was 0.22 nm, which is between the value of fcc Pt(111) and fcc-Ru(111) facets, suggesting the {111} facet of Pt–Ru alloy. **c** XRD spectra. **d** EDS elemental mapping

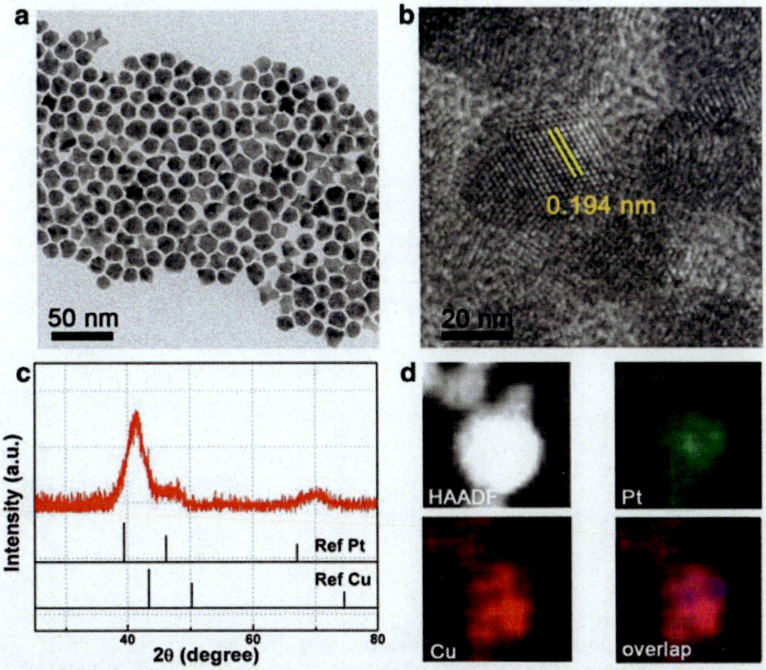

Fig. 3.20 Characterizations of PtCu NPs. **a**, **b** TEM and HRTEM images of PtCu. **c** XRD spectra. The peaks in XRD pattern situated between the typical fcc of Pt and Cu, demonstrating a typical PtCu alloy phase. **d** EDS elemental mapping. As shown, both the Pt and Cu exhibited homogeneous distribution throughout the whole nanostructure

catalysts between the initial sample and that after running the chronopotentiometry test (Fig. 3.34). Notably, Ru kept the same oxidation state after the u–t measurements, considering the possible charge compensation from Pt–Cu to Ru, thereby avoiding the over-oxidation of Ru. To prove the AEM for the generation of molecule oxygen, we further performed in situ attenuated total reflection infrared (ATR-IR) spectroelectro-chemical measurements to capture the OOH intermediate (Figs. 3.17g and 3.35). The recorded peak at approximately 1212 cm^{-1} is assigned to the characteristic vibration adsorption of the surface-adsorbed superoxide (OOH_{ad}) [55]. This indicates that the AEM rather than the LOER mechanism dominated O_2 generation over the Ru_1–Pt_3Cu, which certainly improved its stability under acidic conditions.

Density functional theory calculations on various $Pt_xCu_{4-x}(111)$ ($x = 0, 1, 2, 3, 4$) surfaces were performed to rationalize the observed activity and dissolution resistance. For the dissolution and segregation of Cu under acidic OER conditions, Pt-skin surfaces [56] were used if not stated otherwise, and one-quarter of surface atoms were substituted by Ru to model atomically dispersed Ru_1. Importantly, 0.5 ML oxygen was pre-adsorbed on the surface to simulate the oxidic environment of the Pt_{skin} shell and the Ru atom, as proved by the EXAFS and XANES analyses (Fig. 3.36b–f). The critical minimum overpotential η, required to enable all OER elementary

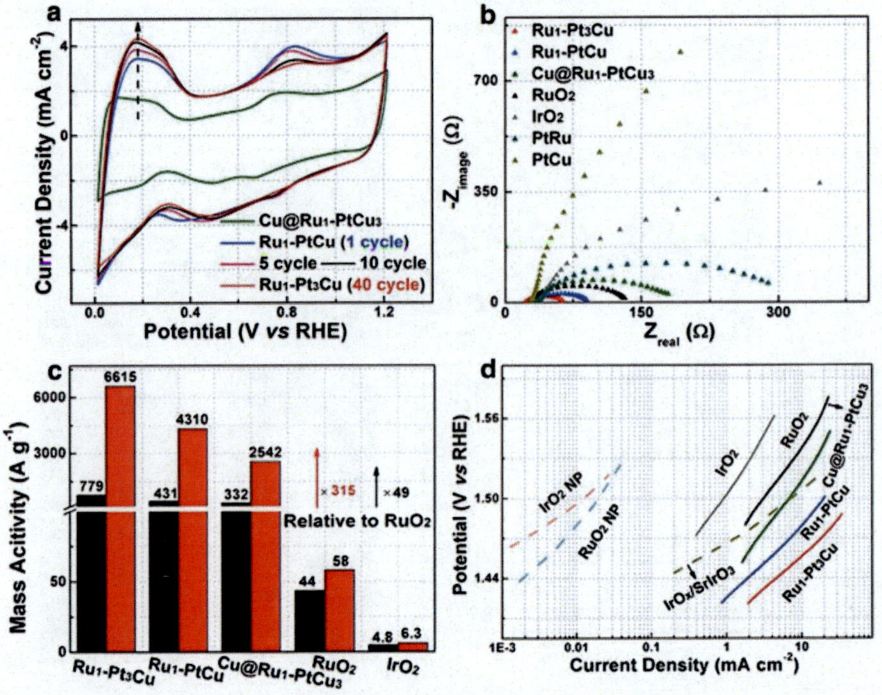

Fig. 3.21 OER electrochemical performance of typical samples. **a** The changes of CV curves for several processes. **b** EIS spectra of different catalysts under the same experimental condition. The data were collected for the electrodes under 1.20 V versus Ag/AgCl. **c** Mass activity of Ru_1–Pt_3Cu, Ru_1–PtCu, Cu@Ru_1–$PtCu_3$, RuO_2 and IrO_2 at $\eta = 280$ mV. **d** Tafel plots. The solid lines were from this work and the dashed line were from literatures [9, 38]

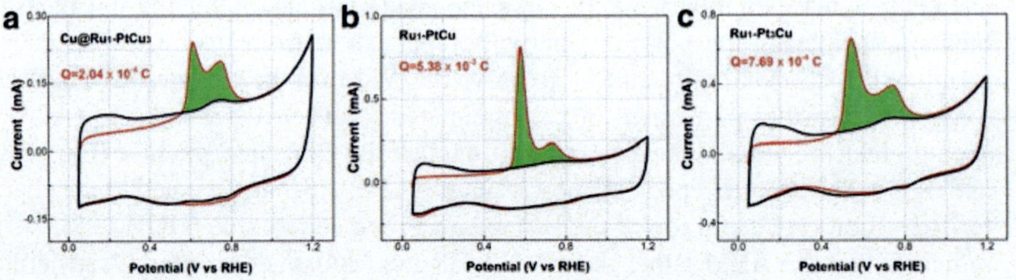

Fig. 3.22 CO stripping curves of three samples. Electrochemical surface characterization of **a** Cu@Ru_1–$PtCu_3$, **b** Ru_1–PtCu and **c** Ru_1–Pt_3Cu by using CO stripping curves [39, 40]

steps of the OER to be exergonic, was investigated by calculating the free energy profile [38, 57, 58]. Specifically, we calculated the corresponding free energies of OH, O and OOH, the key OER intermediates dominating the overall OER activity. A linear scaling relation between the free energies of OOH and OH on atomically dispersed Ru_1 embedded in the alloy surfaces, $\Delta G_{OOH} = 0.93\ \Delta G_{OH} + 3.02 \pm 0.18$

Table 3.5 Comparison of calculated surface areas (ECSAs) obtained from H_{upd} and CO stripping for Cu@Ru_1–$PtCu_3$, Ru_1–PtCu and Ru_1–Pt_3Cu

Electrocatalysts	Cu@Ru_1–$PtCu_3$	Ru_1–PtCu	Ru_1–Pt_3Cu
H adsorption/desorption (m^2/g_{Pt})	36.4	59.1	74.5
CO stripping (m^2/g_{Pt})	35.2	58.3	75.8
ECSA H_{upd}/$ECSA_{CO}$	1.03	1.01	0.98

Table 3.6 Comparison of the overpotential at 10 mA/cm^2 compared to previously reported catalysts in acidic electrolyte

Electrocatalysts	Electrolyte	[a]Overpotential (10 mA/cm^2) (mV)	References
Ru_1–Pt_3Cu	0.1 M $HClO_4$	~220	This work
Ru_1–PtCu	0.1 M $HClO_4$	~250	This work
Cu@Ru_1–$PtCu_3$	0.1 M $HClO_4$	~280	This work
IrO_x/$SrIrO_3$	0.5 M H_2SO_4	~270	[38]
$Cu_{1.5}Mn_{1.5}O_4$:10F	0.5 M H_2SO_4	~320	[43]
Surface-clean 3D Ir	0.1 M $HClO_4$	~270	[44]
Co-IrCu ONC/C	0.1 M $HClO_4$	~293	[45]
Ir–Ni TL	0.05 M H_2SO_4	>370	[46]
IrO_2@RuO_2	0.5 M H_2SO_4	~260	[47]
$Ru_{0.9}I_{r0.1}O_2$	0.5 M H_2SO_4	>270	[48]
r-RuO_2	0.1 M $HClO_4$	>420	[9]
r-IrO_2	0.1 M $HClO_4$	>420	[9]
$IrNiO_x$/meso-ATO-180	0.05 M H_2SO_4	~320	[49]
Dimensional ordered macroporous IrO_2	0.5 M H_2SO_4	~320	[50]
IrO_2/Sb–SnO_2 NW	0.5 M H_2SO_4	~250	[50]
Nanoporous IrO_2	0.5 M H_2SO_4	~282	[51]

[a]Obtained at the current density of 10 mA/cm^2. All potentials were versus to reversible hydrogen electrode (RHE) with *iR*-correction

(Fig. 3.36a), was found and is in good agreement with previous work [12]. Due to the competition between the free energies of the four elementary steps involved [59] (Fig. 3.37), the calculated overpotential η can be represented by a 2D volcano-type surface with respect to the free energy of the O and OH intermediates (Fig. 3.36g). As expected, a too weak binding of adsorbate over the active Ru_1 site (higher ΔG_{OH} and $\Delta G_O - \Delta G_{OH}$) impedes the adsorption of OH and increases the subsequent dehydrogenation barrier. On the other hand, a too strong interaction (lower ΔG_{OH} and $\Delta G_O - \Delta G_{OH}$) is detrimental for formation of the OOH intermediate and subsequent generation of O_2. The best OER activity ($\eta = 0.42$ V) close to the peak of the 2D volcano plot was found on the Ru_1–Pt_3Cu(111) (Fig. 3.36e) surface. In

Table 3.7 Comparisons of the mass activities compared to previously reported catalysts

Electrocatalysts	Electrolyte	[b]Mass activity (A/g) (Ru/Ir)	[a]η (mV)	References
Ru_1–Pt_3Cu	0.1 M $HClO_4$	~6615	280	This work
		~779	250	
Ru_1–PtCu	0.1 M $HClO_4$	~4310	280	This work
		~431	250	
Cu@Ru_1–$PtCu_3$	0.1 M $HClO_4$	~2542	280	This work
		~332	250	
Co-IrCu ONC/C	0.1 M $HClO_4$	~640	300	[45]
Ir–Ni TL	0.05 M H_2SO_4	~498	280	[46]
r-RuO_2 NPs	0.1 M $HClO_4$	~18	250	[9]
r-IrO_2 NPs	0.1 M $HClO_4$	~3.5	250	[9]
$IrNiO_x$/meso-ATO-180	0.05 M H_2SO_4	~90	280	[13]
Dimensional ordered macroporous IrO_2	0.5 M H_2SO_4	~583	370	[49]
IrO_2/Sb–SnO_2 NW	0.5 M H_2SO_4	~816	290	[51]
Nanoporous IrO_2	0.5 M H_2SO_4	~28.5	280	[52]
Ir-ND/ATO	0.05 M H_2SO_4	~70	280	[53]
DO-$IrNi_{3.3}$	0.05 M H_2SO_4	~500	300	[54]
3 nm RuO_2 NPs	0.05 M H_2SO_4	~600	250	[23]

[a]All potentials were versus to reversible hydrogen electrode (RHE) with *iR*-correction. [b]Mass activity was calculated by active species Ru/Ir. In our work, we selected two representative overpotential (250 and 280 mV) to evaluate mass activity of Ru_1–Pt_3Cu, Ru_1–PtCu and Cu@Ru_1–$PtCu_3$

contrast, Ru_1–Cu(111) (Fig. 3.36b), Ru_1–$PtCu_3$(111) (Fig. 3.36c), Ru_1–PtCu(111) (Fig. 3.36d) and Ru_1–Pt(111) (Fig. 3.36f) exhibit overpotentials of 0.82, 0.71 and 0.66 and 0.92 V, respectively, which are less active. Although the trend reproduces the experimental results well, the difference between the calculated (0.42 V) and experimental (0.22 V at 10 mA/cm^2, 0.17 V at 0.1 mA/cm^2) overpotentials for Ru_1–Pt_3Cu remains, which might arise because of the simplified model used in the calculations.

To shed light on the electronic origin of the computationally predicted beneficial reactivity of atomically dispersed Ru_1, the projected density of states (PDOS) of Ru_1 were analyzed and plotted (Fig. 3.36h). The data reveal that, with an increasing Pt to Cu ratio from $PtCu_3$, PtCu, Pt_3Cu to Pt, the PDOS of Ru_1 shifts gradually towards the Fermi level, and the corresponding d-band center $\varepsilon_{Ru\text{-}d}$ shifts up from −3.37 to −2.76 eV. The upshift of the d-band center is understandable because the compressive surface strain with respect to the pristine Pt was gradually released (Fig. 3.36i). As a result, the oxygen adsorption (E_O) is systematically strengthened from −1.63 eV for the pristine Cu (too weak binding) to −2.48 eV for the pristine Pt (too strong binding). This eventually leads to an inverse volcano-type plot of the OER

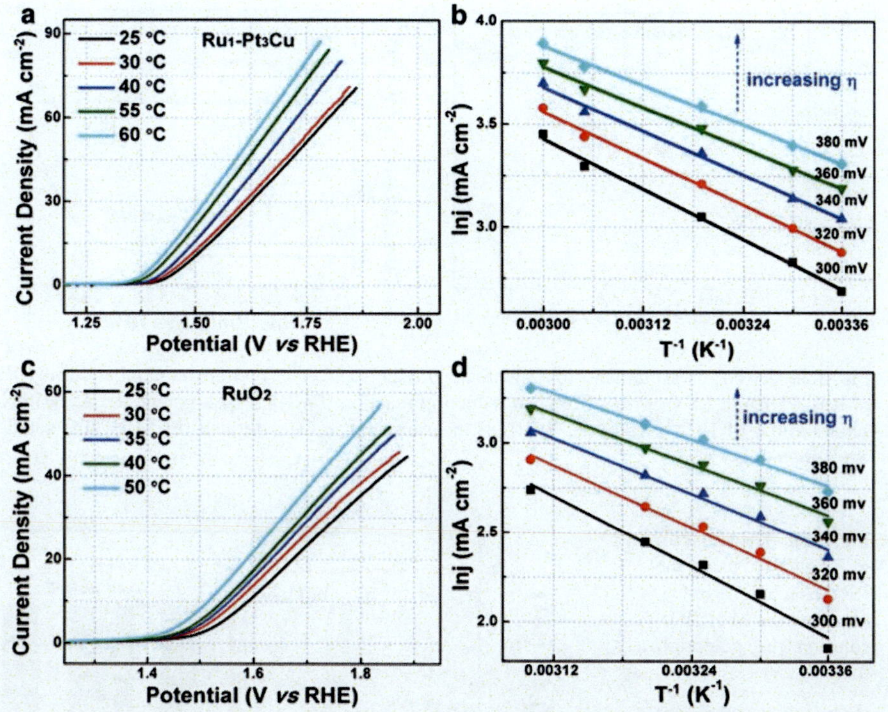

Fig. 3.23 Polarization curves at different temperatures of Ru_1–Pt_3Cu and RuO_2 with a scan rate 10 mV/s in 0.1 M $HClO_4$ solution. **a**, **c** Polarization curves at different temperatures without *iR*-corrected. **b**, **d** Arrhenius plots: semilogarithmic dependence of current density at various overpotentials plotted against inverse temperature. Overpotentials were taken from 300 to 400 mV at an interval of 20 mV

Fig. 3.24 Activation energy at the zero overpotential obtained through trend extrapolation

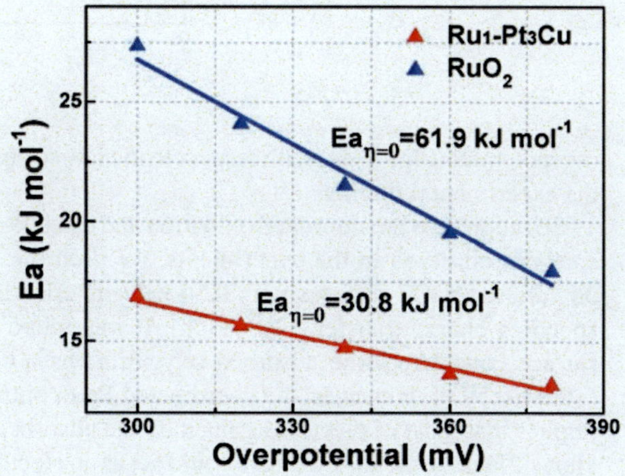

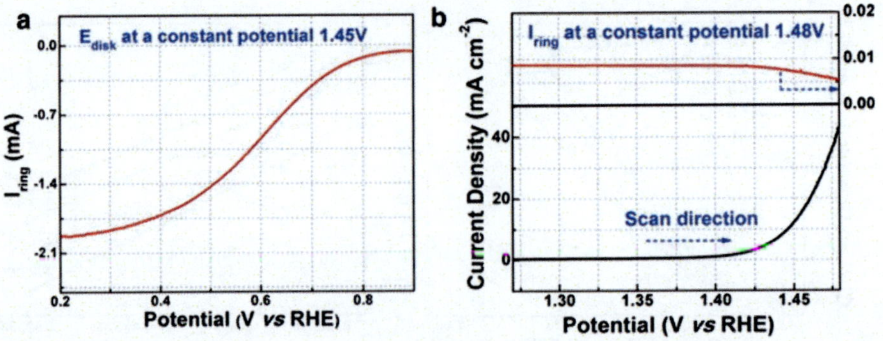

Fig. 3.25 RRDE measurements to evaluate the products and electron transfer number during OER of Ru_1–Pt_3Cu. **a** LSV curve of oxygen reduction reaction (ORR) at disk potential 1.45 V in N_2-saturated 0.1 M $HClO_4$ solution. **b** RRDE measurement in O_2-saturated 0.1 M $HClO_4$ solution (ring potential: 1.48 V)

Fig. 3.26 The ring current of Ru_1–Pt_3Cu using the RRDE technique in O_2 saturated 0.1 M $HClO_4$ solution (ring potential: 0.40 V)

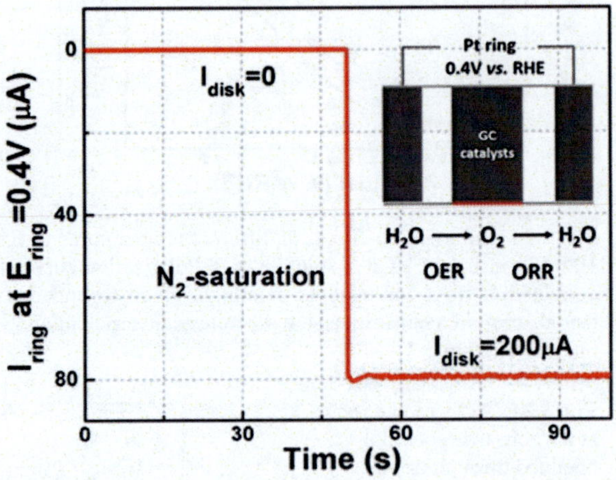

overpotential (Fig. 3.36j) with respect to the lattice constant, and explains excellently the experimental findings.

To rationalize the improved oxidation and dissolution resistance of the Ru_1–Pt_3Cu catalysts observed in the experiments, we used the adsorption of an oxygen atom on various Pt–Cu surfaces with 0.50 ML pre-adsorbed oxygen to probe the corresponding charge transfer (Fig. 3.36k). It was found that, irrespective of the Pt–Cu surface considered, the adsorbed oxygen gains a considerable amount of charge (~0.54–0.59 e). However, the coordinated Ru_1 contributes only ~0.19–0.21 e. This implies that most of electrons gained by the adsorbed oxygen come from the Pt–Cu alloys. Namely, the Pt–Cu alloys could act as an electron reservoir to donate electrons towards reaction intermediates and help prevent the over-oxidation and subsequent dissolution of Ru_1.

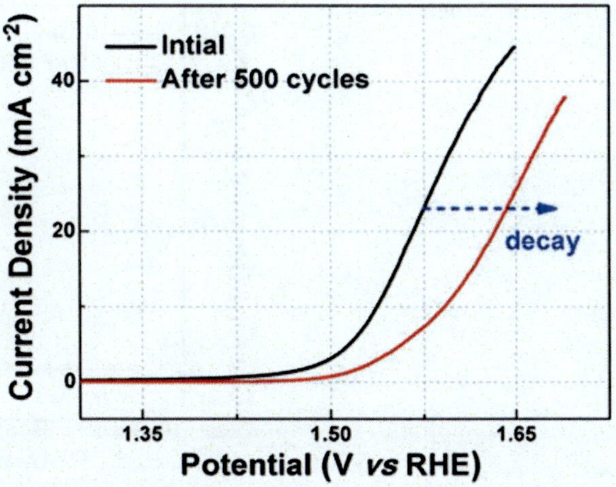

Fig. 3.27 Stability test of commercial RuO_2

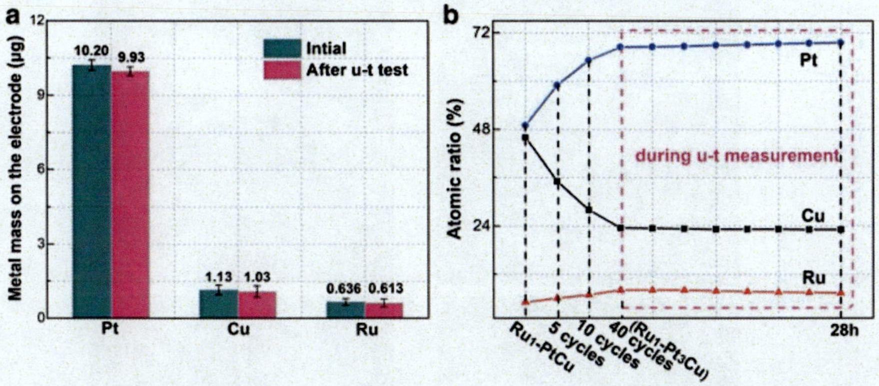

Fig. 3.28 Quantify corrosion by means of inductively coupled mass spectrometry (ICP-MS). **a** The mass change for Pt, Cu and Ru metals after chronopotentiometry tests based on ICP-MS measurements. The error bars show the standard deviation evaluated from five independent measurement. **b** The change of elemental ratio of Ru, Pt and Cu during CV process and u-t measurement

3.4 Conclusion

In summary, we have synthetized a series of Pt–Cu alloys with embedded Ru atoms as an OER electrocatalyst through acid etching and electrochemical leaching. Atomically resolved elemental mapping revealed that the present Ru species was highly atomically dispersed and the $PtCu_x/Pt_{skin}$ core–shell structure was achieved. In-situ XAFS studies demonstrated that the oxidation state of Ru_1 was almost unchanged within the OER catalysis potential range. Accordingly, we have shown how engineering the electronic structure of single Ru sites via compressive strain boosts OER activity by tailoring the binding of oxygenated species close to an optimized level and

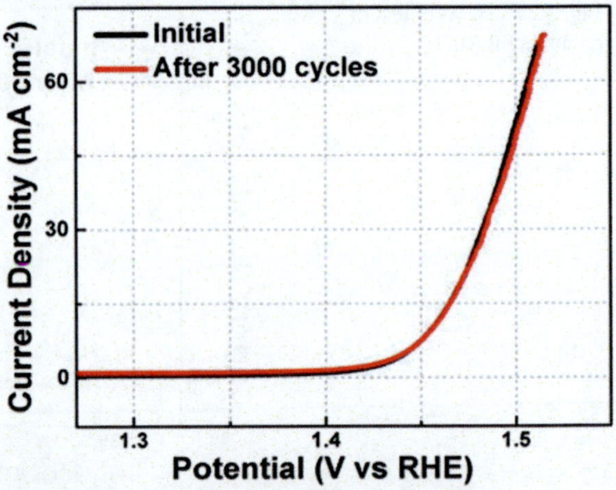

Fig. 3.29 Stability test of $Ru_1–Pt_3Cu$

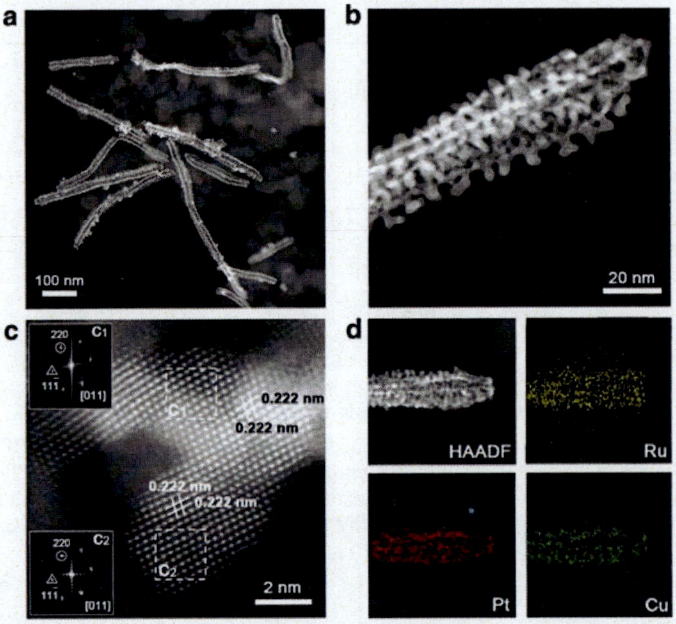

Fig. 3.30 Characterizations of $Ru_1–Pt_3Cu$ after OER reaction. **a**, **b** HAADFSTEM images. As shown, the $Ru_1–Pt_3Cu$ catalyst still retain the initial tubular structure even under such a rigidly corrosive condition. **c** High magnification image and corresponding FFT pattern. **d** EDS elemental mapping

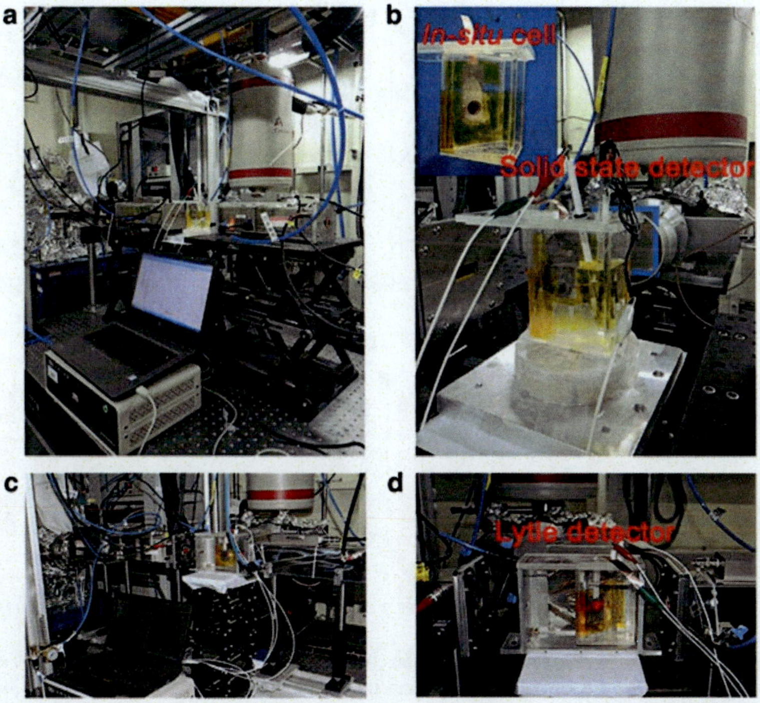

Fig. 3.31 The device and beam path for in-situ XAFS measurement. The experiments were carried out at 14W1 in SSRF (Pt L_3-edge and Cu K-edge, Lytle detector was applied, Ru K-edge, 7-element Ge solid state detector was applied)

suppresses its over-oxidation by effectively preventing the electron over-transferring from Ru to the oxygen-containing ligands.

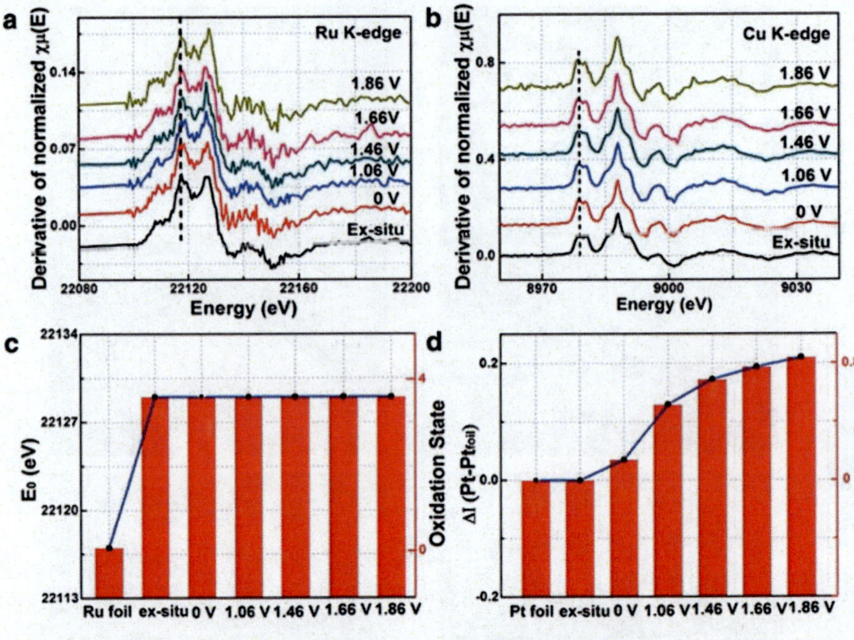

Fig. 3.32 First-derivative to further clarify the change of valance state during OER. First-derivative curves of in-situ XANES of **a** Ru K-edge and **b** Cu K-edge. **c** Histogram dependence on E_0 and average oxidation state of Ru. **d** Histogram dependence on ΔII and average oxidation state of Pt

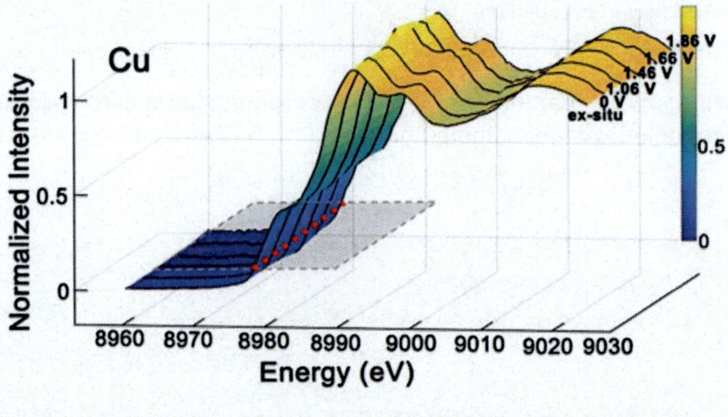

Fig. 3.33 First-derivative to further clarify the change of valance state during OER

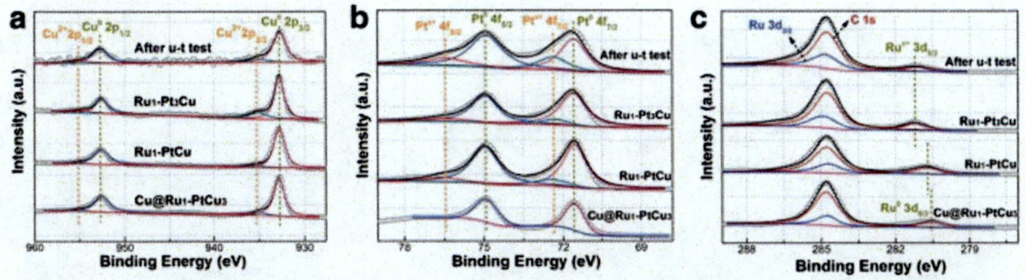

Fig. 3.34 XPS spectra of **a** Pt 4f, **b** Cu 2p and **c** Ru 3d for four different process treatment

Fig. 3.35 Comparison of voltammetry and average current from multipotential application at Ru_1–Pt_3Cu during OER in 0.1 M $HClO_4$ solution

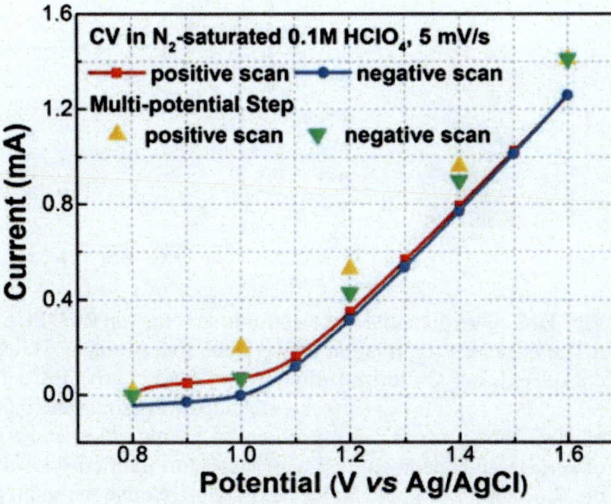

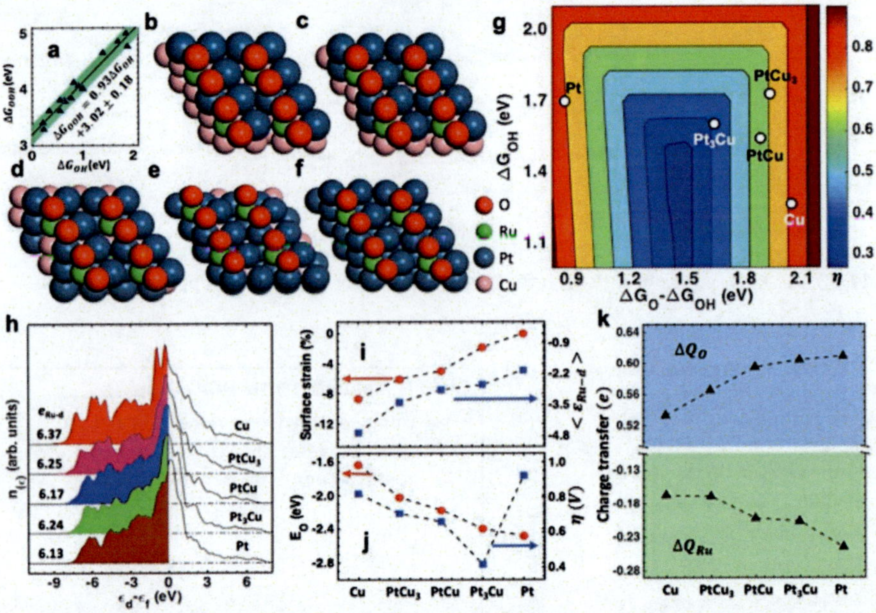

Fig. 3.36 Overpotential and electronic structure on Ru_1–$Pt_xCu_{4-x}(111)$ with pre-adsorbed oxygen. **a** The linear scaling relation between the free energies ΔG of OOH and OH on the alloy surfaces considered. **b–f** Optimized structures of Cu(111) (**b**), $PtCu_3(111)$ (**c**), PtCu(111) (**d**), $Pt_3Cu(111)$ (**e**) and Pt(111) (**f**) with surface-embedded Ru_1 atoms and 0.50 ML pre-adsorbed oxygen. **g** Calculated volcano plot of OER overpotential η with ΔG_{OH} and $\Delta G_O - \Delta G_{OH}$ as descriptors. **h** PDOS of surface-embedded Ru_1 4d with respect to the Fermi level, where $e_{Ru\text{-}d}$ is total 4d electrons of the Ru_1 atom. **i** In-plane lattice contraction relative to the Pt(111) pristine surface (red circles) and corresponding d-band centre $\varepsilon_{Ru\text{-}d}$ of Ru_1 (blue squares). **j** Corresponding adsorption energy E_O of the oxygen atoms (red circles) and η (blue squares). **k** Bader charge ΔQ_O for adsorbed oxygen and variation of Bader charge ΔQ_{Ru} for coordinated Ru_1 on alloy surfaces with 0.50 ML pre-adsorbed oxygen

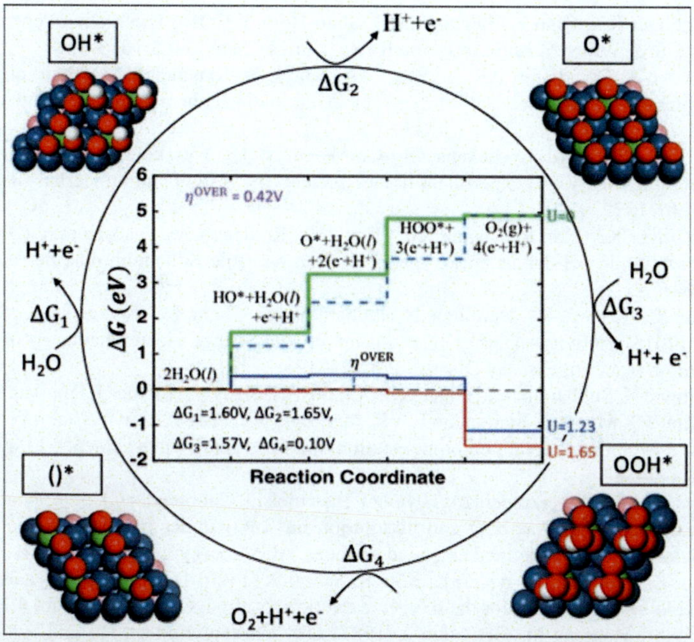

Fig. 3.37 OER mechanism and corresponding Gibbs free energy change and adsorption structures on Ru embedded Pt-skin $Pt_3Cu(111)$

References

1. Chu S, Majumdar A (2012) Opportunities and challenges for a sustainable energy future. Nature 488:294–303
2. Cebolla VL, Memrado L, Vela J, Ferrrando AC, Romero C (1996) American chemical society, division of fuel chemistry, preprints
3. Kanan MW, Nocera DG (2008) In situ formation of an oxygen-evolving catalyst in neutral water containing phosphate and Co^{2+}. Science 321:1072–1075
4. Vojvodic A, Nørskov JK (2011) Optimizing perovskites for the water-splitting reaction. Science 334:1355–1356
5. Yin Q, Tan JM, Besson C, Geletii YV, Musaev DG, Kuznetsov AE, Luo Z, Hardcastle KI, Hill CL (2010) A fast soluble carbon-free molecular water oxidation catalyst based on abundant metals. Science 328:342–345
6. Spoeri C, Kwan JTH, Bonakdarpour A, Wilkinson DP, Strasser P (2017) The stability challenges of oxygen evolving catalysts: towards a common fundamental understanding and mitigation of catalyst degradation. Angew Chem Int Ed 56:5994–6021
7. McCrory CC, Jung S, Ferrer IM, Chatman SM, Peters JC, Jaramillo TF (2015) Benchmarking hydrogen evolving reaction and oxygen evolving reaction electrocatalysts for solar water splitting devices. J Am Chem Soc 137:4347–4357
8. Carmo M, Fritz DL, Mergel J, Stolten D (2013) A comprehensive review on PEM water electrolysis. Int J Hydrogen Energy 38:4901–4934
9. Lee Y, Suntivich J, May KJ, Perry EE, Shao-Horn Y (2012) Synthesis and activities of rutile IrO_2 and RuO_2 nanoparticles for oxygen evolution in acid and alkaline solutions. J Phys Chem Lett 3:399–404

10. Stoerzinger KA, Qiao L, Biegalski MD, Shao-Horn Y (2014) Orientation-dependent oxygen evolution activities of rutile IrO_2 and RuO_2. J Phys Chem Lett 5:1636–1641
11. Cherevko S, Zeradjanin AR, Topalov AA, Kulyk N, Katsounaros I, Mayrhofer KJ (2014) Dissolution of noble metals during oxygen evolution in acidic media. ChemCatChem 6:2219–2223
12. Seh ZW, Kibsgaard J, Dickens CF, Chorkendorff IB, Nørskov JK, Jaramillo TF (2017) Combining theory and experiment in electrocatalysis: insights into materials design. Science 355:146–157
13. Stoerzinger KA, Rao RR, Wang XR, Hong WT, Rouleau CM, Shao-Horn Y (2017) The role of Ru redox in pH-dependent oxygen evolution on rutile ruthenium dioxide surfaces. Chem 2:668–675
14. Chang SH, Connell JG, Danilovic N, Subbaraman R, Chang KC, Stamenkovic VR, Markovic NM (2015) Activity–stability relationship in the surface electrochemistry of the oxygen evolution reaction. Faraday Discuss 176:125–133
15. Danilovic N, Subbaraman R, Chang KC, Chang SH, Kang YJ, Snyder J, Paulikas AP, Strmcnik D, Kim Y-T, Myers D, Stamenkovic VR, Markovic NM (2014) Activity-stability trends for the oxygen evolution reaction on monometallic oxides in acidic environments. J Phys Chem Lett 5:2474–2478
16. Roy C, Rao RR, Stoerzinger KA, Hwang J, Rossmeisl J, Chorkendorff I, Shao-Horn Y, Stephens IE (2018) Trends in activity and dissolution on RuO_2 under oxygen evolution conditions: particles versus well-defined extended surfaces. ACS Energy Lett 3:2045–2051
17. Fabbri E, Habereder A, Waltar K, Kötz R, Schmidt TJ (2014) Developments and perspectives of oxide-based catalysts for the oxygen evolution reaction. Catal Sci Technol 4:3800–3821
18. Wohlfahrt-Mehrens M, Heitbaum J (1987) Oxygen evolution on Ru and RuO_2 electrodes studied using isotope labelling and on-line mass spectrometry. J Electroanal Chem 237:251–260
19. Binninger T, Mohamed R, Waltar K, Fabbri E, Levecque P, Kötz R, Schmidt TJ (2015) Thermodynamic explanation of the universal correlation between oxygen evolution activity and corrosion of oxide catalysts. Sci Rep 5:12167–12172
20. Grimaud A, Diaz-Morales O, Han B, Hong WT, Lee YL, Giordano L, Stoerzinger KA, Koper MT, Shao-Horn Y (2017) Activating lattice oxygen redox reactions in metal oxides to catalyse oxygen evolution. Nat Chem 9:457–465
21. Iwakura C, Hirao K, Tamura H (1977) Anodic evolution of oxygen on ruthenium in acidic solutions. Electrochim Acta 22:329–334
22. Hodnik N, Jovanovič P, Pavlišič A, Jozinović B, Zorko M, Bele M, Šelih VS, Šala M, Hočevar S, Gaberšček M (2015) New insights into corrosion of ruthenium and ruthenium oxide nanoparticles in acidic media. J Phys Chem C 119:10140–10147
23. Paoli EA, Masini F, Frydendal R, Deiana D, Schlaup C, Malizia M, Hansen TW, Horch S, Stephens IE, Chorkendorff I (2015) Oxygen evolution on well-characterized mass-selected Ru and RuO_2 nanoparticles. Chem Sci 6:190–196
24. Rong X, Parolin J, Kolpak AM (2016) A fundamental relationship between reaction mechanism and stability in metal oxide catalysts for oxygen evolution. ACS Catal 6:1153–1158
25. Strasser P (2016) Free electrons to molecular bonds and back: closing the energetic oxygen reduction (ORR)–oxygen evolution (OER) cycle using core–shell nanoelectrocatalysts. Acc Chem Res 49:2658–2668
26. Kötz R, Stucki S, Scherson D, Kolb DM (1984) In-situ identification of RuO_4 as the corrosion product during oxygen evolution on ruthenium in acid media. J Electroanal Chem 172:211–219
27. AlYami NM, LaGrow AP, Joya KS, Hwang J, Katsiev K, Anjum DH, Losovyj Y, Sinatra L, Kim JY, Bakr OM (2016) Tailoring ruthenium exposure to enhance the performance of fcc platinum@ruthenium core–shell electrocatalysts in the oxygen evolution reaction. Phys Chem Chem Phys 18:16169–16178
28. Cui C, Gan L, Heggen M, Rudi S, Strasser P (2013) Compositional segregation in shaped Pt alloy nanoparticles and their structural behaviour during electrocatalysis. Nat Mater 12:765–771

29. Gan L, Cui C, Heggen M, Dionigi F, Rudi S, Strasser P (2014) Element-specific anisotropic growth of shaped platinum alloy nanocrystals. Science 346:1502–1506
30. Strasser P, Koh S, Anniyev T, Greeley J, More K, Yu C, Liu Z, Kaya S, Nordlund D, Ogasawara H, Toney MF, Nilsson A (2010) Lattice-strain control of the activity in dealloyed core–shell fuel cell catalysts. Nat Chem 2:454–460
31. Huang X, Chen Y, Chiu CY, Zhang H, Xu Y, Duan X, Huang Y (2013) A versatile strategy to the selective synthesis of Cu nanocrystals and the in situ conversion to CuRu nanotubes. Nanoscale 5:6284–6290
32. Baletto F, Ferrando R (2005) Structural properties of nanoclusters: energetic, thermodynamic, and kinetic effects. Rev Mod Phys 77:371–422
33. Kyriakou G, Boucher MB, Jewell AD, Lewis EA, Lawton TJ, Baber AE, Tierney HL, Flytzani-Stephanopoulos M, Sykes ECH (2012) Isolated metal atom geometries as a strategy for selective heterogeneous hydrogenations. Science 335:1209–1212
34. Alayoglu S, Nilekar AU, Mavrikakis M, Eichhorn B (2008) Ru–Pt core–shell nanoparticles for preferential oxidation of carbon monoxide in hydrogen. Nat Mater 7:333–338
35. Li J, Yin HM, Li XB, Okunishi E, Shen YL, He J, Tang Z-K, Wang W-X, Yücelen E, Li C, Gong Y, Gu L, Miao S, Liu L-M, Luo J, Ding Y (2017) Surface evolution of a Pt–Pd–Au electrocatalyst for stable oxygen reduction. Nat Energy 2:17111–17119
36. Lytle FW (1976) Determination of d-band occupancy in pure metals and supported catalysts by measurement of the LIII X-ray absorption threshold. J Catal 43:376–379
37. Kau LS, Hodgson KO, Solomon EI (1989) X-ray absorption edge and EXAFS study of the copper sites in zinc oxide methanol synthesis catalysts. J Am Chem Soc 111:7103–7109
38. Seitz LC, Dickens CF, Nishio K, Hikita Y, Montoya J, Doyle A, Kirk C, Vojvodic A, Hwang HY, Norskov JK, Jaramillo TF (2016) A highly active and stable $IrO_x/SrIrO_3$ catalyst for the oxygen evolution reaction. Science 353:1011–1014
39. van der Vliet DF, Wang C, Li D, Paulikas AP, Greeley J, Rankin RB, Strmcnik D, Tripkovic D, Markovic NM, Stamenkovic VR (2012) Unique electrochemical adsorption properties of Pt-skin surfaces. Angew Chem 124:3193–3196
40. Chen C, Kang Y, Huo Z, Zhu Z, Huang W, Xin HL, Snyder JD, Li D, Herron JA, Mavrikakis M, Chi M, More KL, Li Y, Markovic NM, Somorjai GA, Yang P, Stamenkovic VR (2014) Highly crystalline multimetallic nanoframes with three-dimensional electrocatalytic surfaces. Science 343:1339–1343
41. Denton AR, Ashcrof NW (1991) Vegard's law. Phys Rev A 43:3161–3164
42. Shi Y, Wang J, Wang C, Zhai TT, Bao WJ, Xu JJ, Xia XH, Chen HY (2015) Hot electron of Au nanorods activates the electrocatalysis of hydrogen evolution on MoS_2 nanosheets. J Am Chem Soc 137:7365–7370
43. Patel PP, Datta MK, Velikokhatnyi OI, Kuruba R, Damodaran K, Jampani P, Gattu B, Shanthi PM, Damle SS, Kumta PN (2016) Noble metal-free bifunctional oxygen evolution and oxygen reduction acidic media electro-catalysts. Sci Rep 6:28367
44. Pi Y, Zhang N, Guo S, Guo J, Huang X (2016) Ultrathin laminar Ir superstructure as highly efficient oxygen evolution electrocatalyst in broad pH range. Nano Lett 16:4424–4430
45. Kwon T, Hwang H, Sa YJ, Park J, Baik H, Joo SH, Lee K (2017) Cobalt assisted synthesis of IrCu hollow octahedral nanocages as highly active electrocatalysts toward oxygen evolution reaction. Adv Funct Mater 27:1604688
46. Lim J, Yang S, Kim C, Roh CW, Kwon Y, Kim YT, Lee H (2016) Shaped Ir-Ni bimetallic nanoparticles for minimizing Ir utilization in oxygen evolution reaction. Chem Commun 52:5641–5644
47. Audichon T, Napporn TW, Canaff C, Morais C, Comminges C, Kokoh KB (2016) IrO_2 coated on RuO_2 as efficient and stable electroactive nanocatalysts for electrochemical water splitting. J Phys Chem C 120:2562–2573
48. Audichon T, Mayousse E, Morisset S, Morais C, Comminges C, Napporn TW, Kokoh KB (2014) Electroactivity of RuO_2–IrO_2 mixed nanocatalysts toward the oxygen evolution reaction in a water electrolyzer supplied by a solar profile. Int J Hydrogen Energy 39:e16796

49. Nong HN, Oh HS, Reier T, Willinger E, Willinger MG, Petkov V, Teschner D, Strasser P (2015) Oxide-supported $IrNiO_x$ core-shell particles as efficient, cost-effective, and stable catalysts for electrochemical water splitting. Angew Chem Int Ed 54:2975–2979
50. Hu W, Wang Y, Hu X, Zhou Y, Chen S (2012) Three-dimensional ordered macroporous IrO_2 as electrocatalyst for oxygen evolution reaction in acidic medium. J Mater Chem 22:6010–6016
51. Liu G, Xu J, Wang Y, Wang X (2015) An oxygen evolution catalyst on an antimony doped tin oxide nanowire structured support for proton exchange membrane liquid water electrolysis. J Mater Chem A 3:20791–20800
52. Li G, Li S, Xiao M, Ge J, Liu C, Xing W (2017) Nanoporous IrO_2 catalyst with enhanced activity and durability for water oxidation owing to its micro/meso-pore structure. Nanoscale 9:291–9298
53. Oh H-S, Nong HN, Reier T, Gliech M, Strasser P (2015) Oxide-supported Ir nanodendrites with high activity and durability for the oxygen evolution reaction in acid PEM water electrolyzers. Chem Sci 6:3321–3328
54. Nong HN, Gan L, Willinger E, Teschner D, Strasser P (2014) IrO_x core-shell nanocatalysts for cost- and energy-efficient electrochemical water splitting. Chem Sci 5:2955–2963
55. Nayak S, McPherson IJ, Vincent KA (2018) Adsorbed intermediates in oxygen reduction on platinum nanoparticles observed by in situ IR spectroscopy. Angew Chem 130:13037–13040
56. Stamenkovic VR, Fowler B, Mun BS, Wang G, Ross PN, Lucas CA, Marković NM (2007) Improved oxygen reduction activity on Pt_3Ni (111) via increased surface site availability. Science 315:493–497
57. Zhang B, Zheng X, Voznyy O, Comin R, Bajdich M, García-Melchor M, Han L, Xu J, Liu M, Zheng L, García de Arquer FP, Dinh CT, Fan F, Yuan M, Yassitepe E, Chen N, Regier T, Liu P, Li Y, Luna PD, Janmohamed A, Xin HL, Yang H, Vojvodic A, Sargent EH (2016) Homogeneously dispersed multimetal oxygen-evolving catalysts. Science 352:333–337
58. Suntivich J, May KJ, Gasteiger HA, Goodenough JB, Shao-Horn Y (2011) A perovskite oxide optimized for oxygen evolution catalysis from molecular orbital principles. Science 334:1383–1385
59. Bajdich M, García-Mota M, Vojvodic A, Nørskov JK, Bell AT (2013) Theoretical investigation of the activity of cobalt oxides for the electrochemical oxidation of water. J Am Chem Soc 135:13521–13530

Chapter 4
Engineering the Electronic Structure of Submonolayer Pt on Intermetallic Pd_3Pb via Charge Transfer Boosts the Hydrogen Evolution Reaction

4.1 Introduction

Hydrogen as a clean energy carrier in fuel cells is industrially produced from carbon feedstocks via various processes [1–4], which are environmentally unfriendly because of the release of CO_2. The electrochemical hydrogen evolution reaction (HER) in water electrolysis is an alternative way to produce H_2 [5–7]. The HER involving the reduction of H^+ and desorption of H_2 plays a key role in the water splitting process [8–10]. The design of highly active and stable electrocatalysts is a prerequisite for the HER process. Currently, this process is still largely limited to the use of precious metals such as Pt, especially for industrial requirements [11, 12]. This largely increases the cost of electrolyzers and prevents their large-scale applications.

Numerous efforts to lower the cost of these metals via tuning the compositions and morphologies by increasing their atomic efficiency and intrinsic activity have been reported [13–16]. Since the surface atoms always play a dominant role in the catalytic process, deposition of a precious metal monolayer or submonolayer on a nonprecious substrate is an ideal strategy to extremely enhance the mass activity [17, 18]. However, it is worth noting that stability problems of catalyst would apparently emerge when the as-grown layer is reduced to the atomic scale [19, 20]. To solve this problem, choosing a chemically stable substrate that can resist acid etching and finding an effective surface engineering tool to greatly enhance the interaction between substrate and as-grown atomic overlayer are crucial to the catalyst design [21, 22].

Y. Yao, *Controllable Synthesis and Atomic Scale Regulation of Noble Metal Catalysts*, Springer Theses, https://doi.org/10.1007/978-981-19-0205-5_4

4.2 Experiments

4.2.1 Materials

Name	Formula	Purities	Manufacture
Palladium acetylacetonate	$Pd(acac)_2$	A.R.	Alfa Aesar
Platinum acetylacetonate	$Pt(acac)_2$	A.R.	Alfa Aesar
Lead acetylacetonate	$Pb(acac)_2$	A.R.	Alfa Aesar
Copper bromide	$CuBr_2$	A.R.	Sinopharm
Platinum carbon	Pt/C	20% Pt	Alfa Aesar
Ethanol	CH_3OH	A.R.	Sinopharm
Hexane	C_6H_{14}	A.R.	Sinopharm
Oleylamine	OAm	A.R.	Sigma-Aldrich
Octadecene	ODE	A.R.	Sigma-Aldrich
5% Nafion	Nafion	5%	Sigma-Aldrich
Nitrogen	N_2	99.99%	Nanjing Special Gas Corp

The reagents were directly used without further purities

4.2.2 Methods

Preparation of Pd_3Pb nanoplates. In a typical procedure for synthesizing Pd_3Pb nanoplates, $Pb(acac)_2$ (8 mg) and $Pd(acac)_2$ (7.5 mg) and NH_4Br (10 mg) were dissolved in 1 ml ethanol and 3 ml oleylamine (OAm). The mixture was sonicated for 40 min. The as-obtained suspension solution was slowly injected to a vial containing 3 ml 1-octadecylene (ODE), which was preheated at 120 °C for 10 min with vigorously stirring and then the temperature was changed from 120 to 200 °C. The reaction lasts 6 h. After being cooled down to room temperature, the products were precipitated by ethanol, separated via centrifugation and further purified by an ethanol-hexane mixture for three times.

Preparation of Pd_3Pb alloy. In a typical procedure for synthesizing Pd_3Pb alloy, $Pb(acac)_2$ (8 mg) and $Pd(acac)_2$ (20 mg) were dissolved in 4 ml oleylamine (OAm) and 1 ml 1-octadecylene (ODE). The mixture was sonicated for 30 min at room temperature. The homogeneous mixture was then heated in an oil bath at 180 °C and maintained at this temperature for 1.5 h. Finally, the final product was collected by centrifugation and washed with ethanol and cyclohexane several times.

Preparation of AL-Pt/Pd_3Pb nanoplates. In a typical procedure, 40 μl, 2 mg/ml of $Pt(acac)_2$ dissolved in oleylamine (OAm) was added to the as-prepared Pd_3Pb seeds, which were dissolved in 2.5 and 2.5 ml ODE mixed solution and then the temperature was changed from 120 to 200 °C followed by 6 h thermal treating at 200 °C. Finally, the products were precipitated by ethanol, separated via centrifugation and further purified by an ethanol-hexane mixture for several times. For the synthesis of Pt/Pd_3Pb alloy nanocrystals, the same conditions as for AL-Pt/Pd_3Pb nanoplates were used.

Preparation of Pt nanoparticles. In a typical procedure for synthesizing Pt nanoparticles, $Pt(acac)_2$ (7.8 mg) were dissolved in 1 ml ethanol and 3 ml oleylamine (OAm). The mixture was sonicated for 20 min. The as-obtained suspension solution was slowly injected to a vial containing 3 ml 1-octadecylene (ODE), which was preheated at 120 °C for 10 min with vigorously stirring and then the temperature was changed from 120 to 200 °C. The reaction lasts 6 h. After being cooled down to room temperature, the products were precipitated by ethanol, separated via centrifugation and further purified by an ethanol-hexane mixture for three times.

4.2.3 Characterizations

Powder X-ray diffraction patterns (PXRD) of samples were recorded using a Rigaku Miniflex-600 with a Cu Kα radiation (Cu Kα, $\lambda = 0.15406$ nm, 40 kV and 15 mA). The morphologies were characterized by TEM (Hitachi-7700, 100 kV). The high-angle annular dark-field scanning transmission electron microscope (HAADF-STEM) images were collected on a JEOL JEM-ARM200F TEM/STEM with a spherical aberration corrector working at 200 kV. Elemental analysis of Pd, Pb, and Pt in the samples was detected by an Optima 7300 DV inductively coupled plasma mass spectrometry (ICP-MS) and energy dispersive X-ray spectroscopy (EDS) conducted on JSM-6700F. The X-ray photoelectron spectroscopy (XPS) was collected on scanning X-ray microprobe (PHI 5000 Verasa, ULAC-PHI, Inc.) using Al Kα radiation and the C1s peak at 284.6 eV as internal standard.

XAFS measurements. The X-ray absorption fine structure spectra data (Pt L_3-edge, Pb L_2-edge and Pd K-edge) were collected at 1W1B station in Beijing Synchrotron Radiation Facility (BSRF, operated at 2.5 GeV with a maximum current of 250 mA) and BL 14W1 station in Shanghai Synchrotron Radiation Facility (SSRF, operated at 3.5 GeV with a maximum current of 250 mA), respectively. All samples were pelletized as disks of 13 mm diameter with 1 mm thickness using graphite powder as a binder.

4.2.4 Electrochemical Measurements

Electrochemical measurements for HER. Electrochemical test was conducted on a CHI 760E electrochemical work station (Chenhua Inc., Shanghai, China) with a three-electrode system in 0.5 M H_2SO_4 aqueous solution. The Ag/AgCl (3 M KCl) electrode was used as a reference electrode and a graphite rod was used as a counter electrode. A constant H_2 flow was maintained in the headspace of the electrolyte during the whole experiment. The synthesized Pd_3Pb and AL-Pt/Pd_3Pb were dispersed onto carbon black (Cabot, Vulcan XC-72) with a metal loading of ~20%. Then, the catalysts were subjected to a 400 °C annealing process for 12 h in argon to remove organic surfactants. The turbid catalyst ink was prepared by mixing 4 mg of the catalyst in 1 ml solution containing 960 μl of ethanol and 40 μl of 5% Nafion solution, followed by ultrasonication for 30 min. Similarly, the catalyst inks for Pd_3Pb and commercial Pt/C samples were prepared by the same method as above.

Next, a certain volume of the catalyst ink was carefully dropped onto a 5 mm diameter glassy carbon (GC) disk electrode. The catalyst layer was allowed to dry under room temperature before an electrochemical measurement. In 0.5 M H_2SO_4, the precious catalyst loading was 40.8 $\mu g_{Pt+Pd}/cm^2$ and the loading of Pt was 1.6 $\mu g_{Pt}/cm^2$ (based on ICP-MS). Electrochemical impedance spectroscopy (EIS) measurements were conducted at −0.22 V (vs. Ag/AgCl) on a rotation electrode under 1600 rpm. The amplitude of the sinusoidal wave was 5 mV, and the frequency scan range was from 100 kHz to 0.1 Hz. All experiments were performed at ambient temperature (25 °C) and all potentials were referenced to the RHE (reversible hydrogen electrode) and conducted with iR-corrected calibration unless noted.

The accelerated durability tests of the catalysts were performed in the H_2-saturated 0.5 M H_2SO_4 electrolyte at room temperature (25 °C) by applying potential cycling between −0.1 and −0.5 V versus Ag/AgCl at a sweep rate of 50 mV/s for 10,000 cycles. Chronoamperometry measurement was obtained at a potential of −14 mV for 25 h.

4.3 Results and Discussion

An intermetallic Pd_3Pb nanoplate was first selected as the substrate because of its excellent chemical resistance to acidic corrosion, which results from its stable and strictly ordered atomic arrangement [23, 24]. First, the Pd_3Pb nanoplates were successfully prepared, and the X-ray diffraction (XRD) pattern (Fig. 4.1) indicated that the intermetallic phase was obtained. Low-magnification high-angle annular dark-field scanning transmission electron microscopy (HAADF-STEM) (Fig. 4.2a) and TEM (Fig. 4.3) images showed that the Pd_3Pb nanoplates presented a square morphology enclosed by {100} facets, which was further confirmed by the electron diffraction pattern (Fig. 4.4) and fast Fourier transform (FFT) pattern (Fig. 4.2b). According to the atomic force microscopy (AFM) image (Fig. 4.5), the thickness of the Pd_3Pb nanoplates was estimated to be 7 nm and the height to width ratio was about 0.17. The atomic resolution HAADFSTEM image (Fig. 4.2b) shows a characteristic lattice spacing of 0.207 nm, which was assigned to the {200} facets of Pd_3Pb. EDS mapping (Fig. 4.2c) revealed that Pd and Pb were distributed uniformly, further demonstrating the intermetallic structure. Subsequently, submonolayer Pt

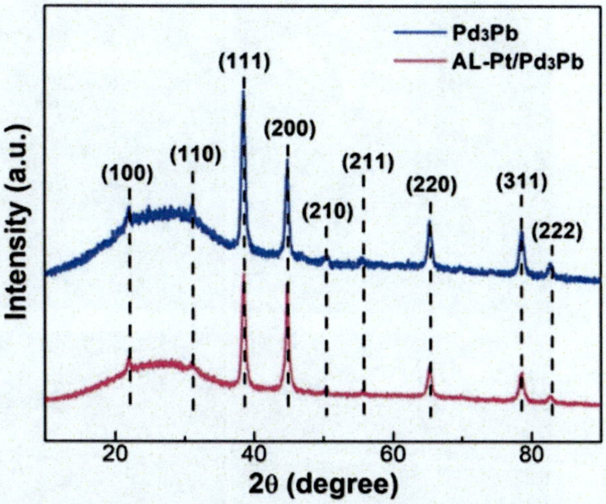

Fig. 4.1 XRD diffraction patterns of Pd_3Pb and AL-Pt/Pd_3Pb

was controllably deposited on Pd_3Pb to obtain atomic-layer Pt/Pd_3Pb (AL-Pt/Pd_3Pb) (Fig. 4.2d). It is reasonable that there were no new peaks assigned to Pt observed in the XRD pattern of AL-Pt/Pd_3Pb (Fig. 4.1) compared with Pd_3Pb because the small proportion of Pt was below the detection limit of XRD.

Despite the successful deposition of a submonolayer of Pt (Fig. 4.6), the lattice spacing of 0.205 nm (Fig. 4.2e) was still assigned to Pd_3Pb {200} facets because the substrate contributed the vast majority of the contrast degree. Since the brightness of an atom column in a HAADF-STEM image is dependent on the atomic number [25, 26], the brightness level of atoms should be Pb (82) > Pt (78) > Pd (46). As shown in Figs. 4.2b and 4.7, the outermost surface of Pd_3Pb was terminated by Pd_1Pb_1 with a PdPbPdPb alignment, and a periodic light and dark contrast arrangement was observed. After the deposition of Pt, an obvious lighter layer covering the surface of Pd_3Pb was observed, and the periodic light and dark arrangement disappeared (Fig. 4.2e). Hence, it is reasonable to conclude that the outermost layer of AL-Pt/Pd_3Pb was the atomic Pt layer. To further confirm the distribution, line intensity profiles along the yellow dashed arrow direction in (Fig. 4.2e) were measured. As shown, the brightness of the outermost Pt atoms was clearly between those of Pb and Pd (Fig. 4.2e_1, e_2). EDS mapping (Fig. 4.2f) clearly displayed the Pd_3Pb core and the thin Pt shell, also confirming the outermost distribution of the Pt layer. Since X-ray photoelectron spectroscopy (XPS) is an effective tool to reveal the electronic structure at the surface of catalysts [27], XPS measurements were performed (Fig. 4.8). Notably, the main peaks of Pb 4f in AL-Pt/Pd_3Pb shifted to higher binding energy compared with Pb in Pd_3Pb, along with a 0.2 eV left shift (Fig. 4.9a). A similar peak shift to higher binding energy (0.15 eV left shift) was also observed for Pd 3d of AL-Pt/Pd_3Pb in comparison with Pd in Pd_3Pb (Fig. 4.10b). As a contrast, the Pt 4f XPS spectrum of AL-Pt/Pd_3Pb (Fig. 4.9c) shifted to lower binding energy (0.3 eV right shift) in comparison with Pt nanoparticles prepared using similar conditions as

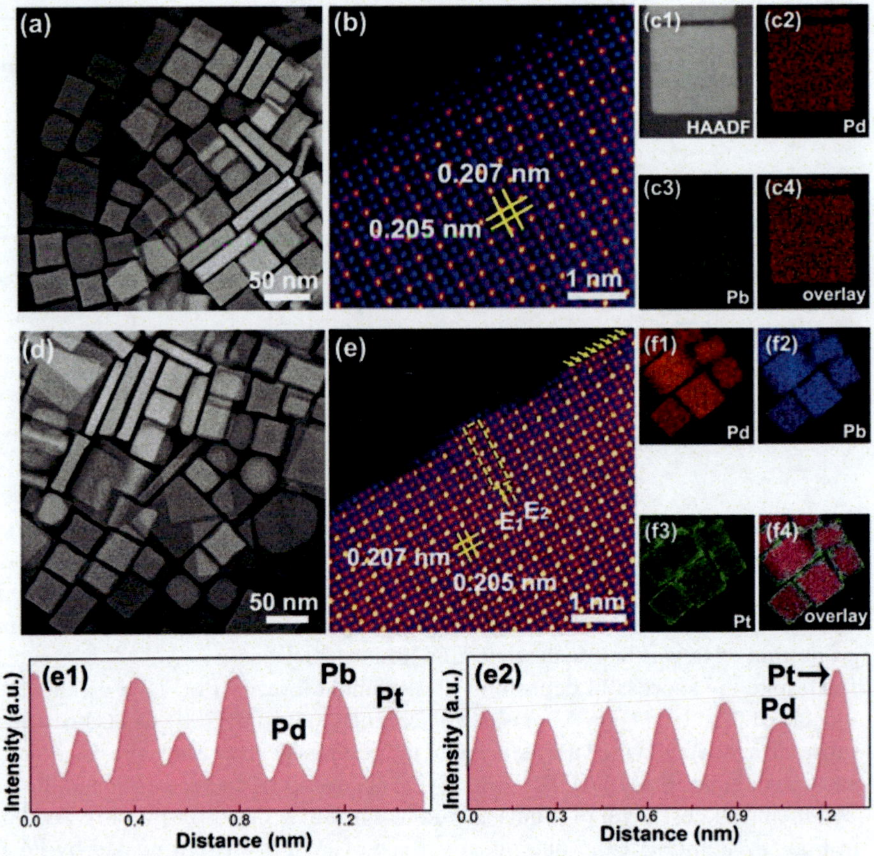

Fig. 4.2 **a** HAADF-STEM image, **b** atomic-resolution HAADFSTEM image and FFT pattern, and **c** EDS mapping of Pd_3Pb. **d** HAADF-STEM and **e** atomic-resolution HAADF-STEM images of AL-Pt/Pd_3Pb. $\mathbf{e}_1$, $\mathbf{e}_2$ Line intensity profiles taken along the yellow dashed arrow direction in **e**. **f** EDS mapping of AL-Pt/Pd_3Pb

for Pd_3Pb (Fig. 4.10). These results underlined the electron transfer from the Pd_3Pb substrate to the Pt layer.

Next, to further explore the local structure of AL-Pt/Pd_3Pb, the X-ray absorption fine structure (XAFS) was measured. The Fourier transform (FT) of the extended XAFS (EXAFS) of the Pd K-edge was almost coincident for AL-Pt/Pd_3Pb and Pd_3Pb (Fig. 4.9d), indicating the similar coordination environments of Pd, and the strong peak at 2.65 Å was attributed to Pd–Pb coordination. The similarity of the Pb coordination environments in Pd_3Pb and AL-Pt/Pd_3Pb was also evidenced by the almost identical curves for the Pb L_2-edge (Fig. 4.11). In the EXAFS spectrum of AL-Pt/Pd_3Pb, the Pt L_3-edge exhibited a peak at 2.17 Å belonging to the scattering of Pt–Pd and a peak at 2.64 Å corresponding to Pt–Pt/Pb coordination (Fig. 4.9e). On the basis of the least-squares EXAFS fitting (Figs. 4.12 and 4.13), the Pt coordination

Fig. 4.3 TEM image of Pd_3Pb nanoplates

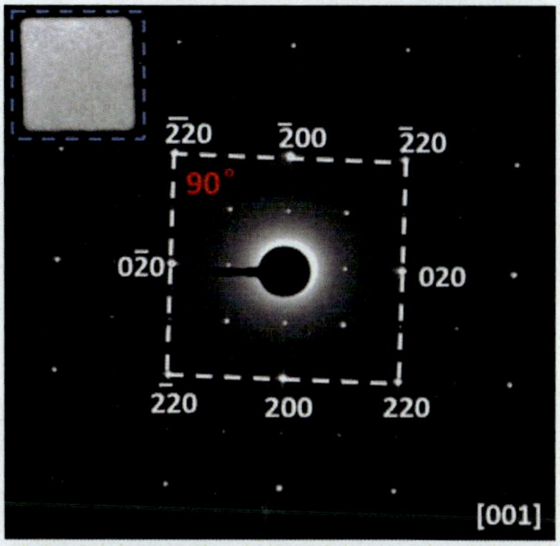

Fig. 4.4 Electron diffraction of a single Pd_3Pb nanoplate

number was just 8.1 (Tables 4.1 and 4.2), which is far less than the saturated coordination number of 12 in Pt foil. This result reinforced the existence of submonolayer Pt at the surface of Pd_3Pb. The X-ray absorption near-edge structure (XANES) shows that the white line intensity of Pt for AL-Pt/Pd_3Pb was below the intensity of Pt foil and commercial Pt/C (Fig. 4.9f). This indicates that Pt exhibited partially negative valence, which resulted from the electron donation from the Pd_3Pb substrate to Pt. For Pb and Pd, the XANES analysis is shown in Figs. 4.14 and 4.15.

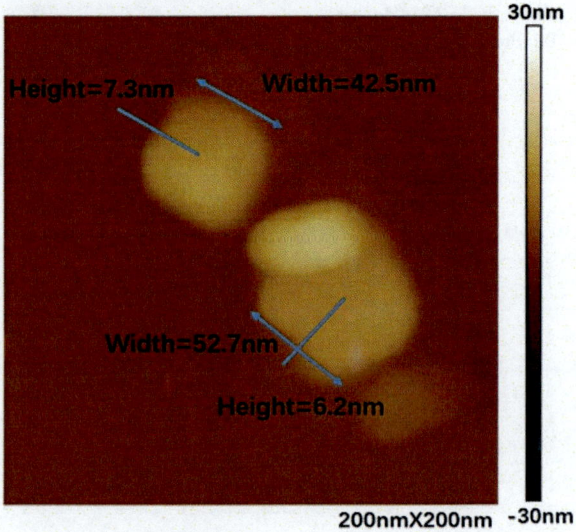

Fig. 4.5 AFM image of Pd_3Pb nanoplates

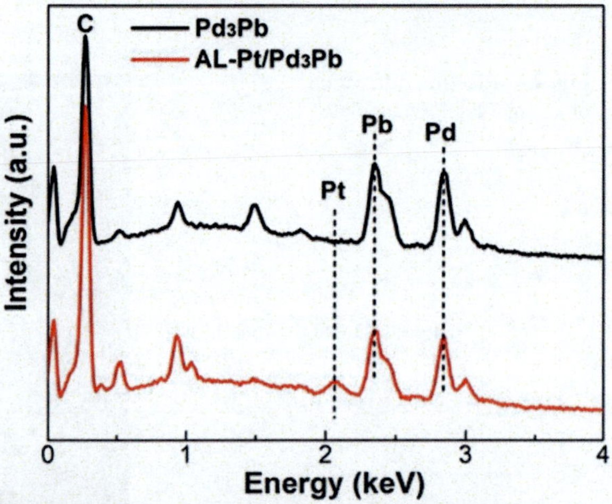

Fig. 4.6 EDS spectra of Pd_3Pb and AL-Pt/Pd_3Pb nanoplates

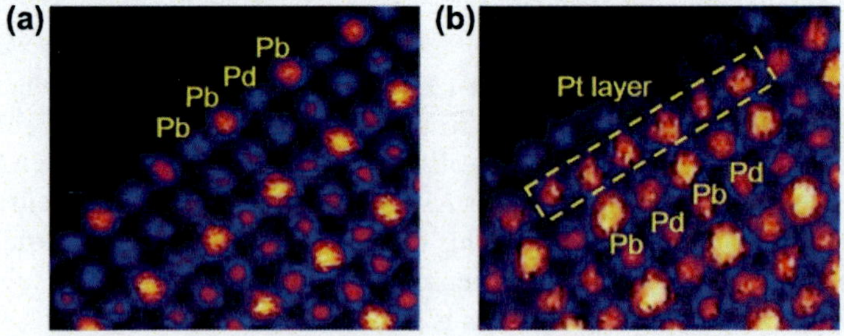

Fig. 4.7 A zoomed view of the surface structure of Fig. 4.2b, e

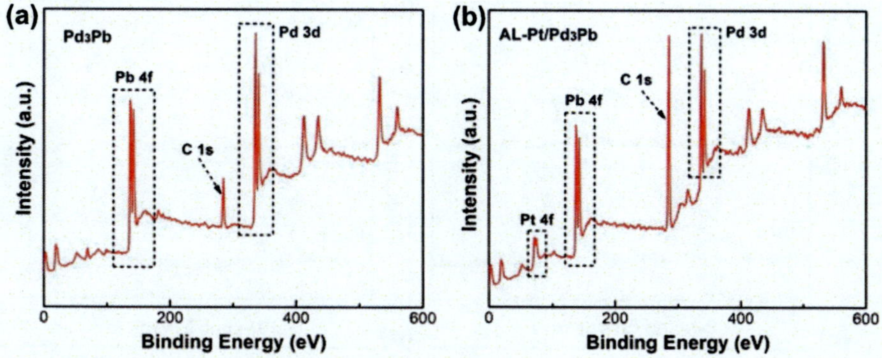

Fig. 4.8 Survey spectra of Pd_3Pb and AL-Pt/Pd_3Pb

Before electrocatalysis measurements, the catalysts underwent an annealing process at 400 °C for 12 h in an argon atmosphere to remove their covering surfactants (Fig. 4.16) [28]. To evaluate the HER activity of AL-Pt/Pd_3Pb, linear sweep voltammetry (LSV) was performed. As shown in Figs. 4.17a and 4.18, AL-Pt/Pd_3Pb possessed a smaller onset potential than Pt/C, Pd_3Pb, and Pt/Pd_3Pb alloy. In addition, we synthesized Pd_3Pb with different numbers of Pt layers and compared their acidic HER performance (Fig. 4.19 and Table 4.3). Interestingly, AL-Pt/Pd_3Pb showed superior HER activity relative to Pd_3Pb with two Pt layers or three to four Pt layers. Besides the onset potential, only 13.8 mV (with iR correction based on Fig. 4.20) was needed to reach 10 mA/cm^2 for AL-Pt/Pd_3Pb, which was much lower than for Pt/C (30 mV) and Pd_3Pb (296 mV). Next, we used the Tafel slope to evaluate the kinetics of samples during the HER (Fig. 4.17b). Remarkably, the Tafel slope for AL-Pt/Pd_3Pb was just 18 mV/dec, which was much lower than that of Pt/C (30 mV/dec), suggesting faster kinetics of the reaction process. For comparison with reported precious metal HER catalysts in acidic media (using the overpotential at 10 mA/cm^2 and the Tafel slope for comparison), AL-Pt/Pd_3Pb presented almost the smallest overpotential and Tafel slope (Fig. 4.17c and Table 4.4). Besides the activities, the stabilities of catalysts should also be taken into full consideration. After a 10,000 cycles electrochemical accelerated durability test (ADT), Pt/C exhibited 46% decay at 50 mA/cm^2, while AL-Pt/Pd_3Pb showed negligible activity decay (Figs. 4.17d and 4.21). In addition, during a continuous 25 h chronoamperometry test at a potential of −14 mV, AL-Pt/Pd_3Pb also showed excellent stability (Fig. 4.17e). At −0.05 V, the mass activity of AL-Pt/Pd_3Pb could reach to 7834 A/g (normalized to Pt and Pd), which was 5.3 times higher than that of Pt/C (1486 A/g) (Fig. 4.17f). When normalized only to Pt, the mass activity of AL-Pt/Pd_3Pb actually reached 22,750 A/g, which is 15-fold higher than that of Pt/C. This was attributed to the maximized utilization of submonolayer Pt atoms on the surface of Pd_3Pb. The atomic-resolution HAADF-STEM image of the recycled sample (Fig. 4.22) verified that the submonolayer Pt was still maintained after the durability test. In addition, the excellent stability of

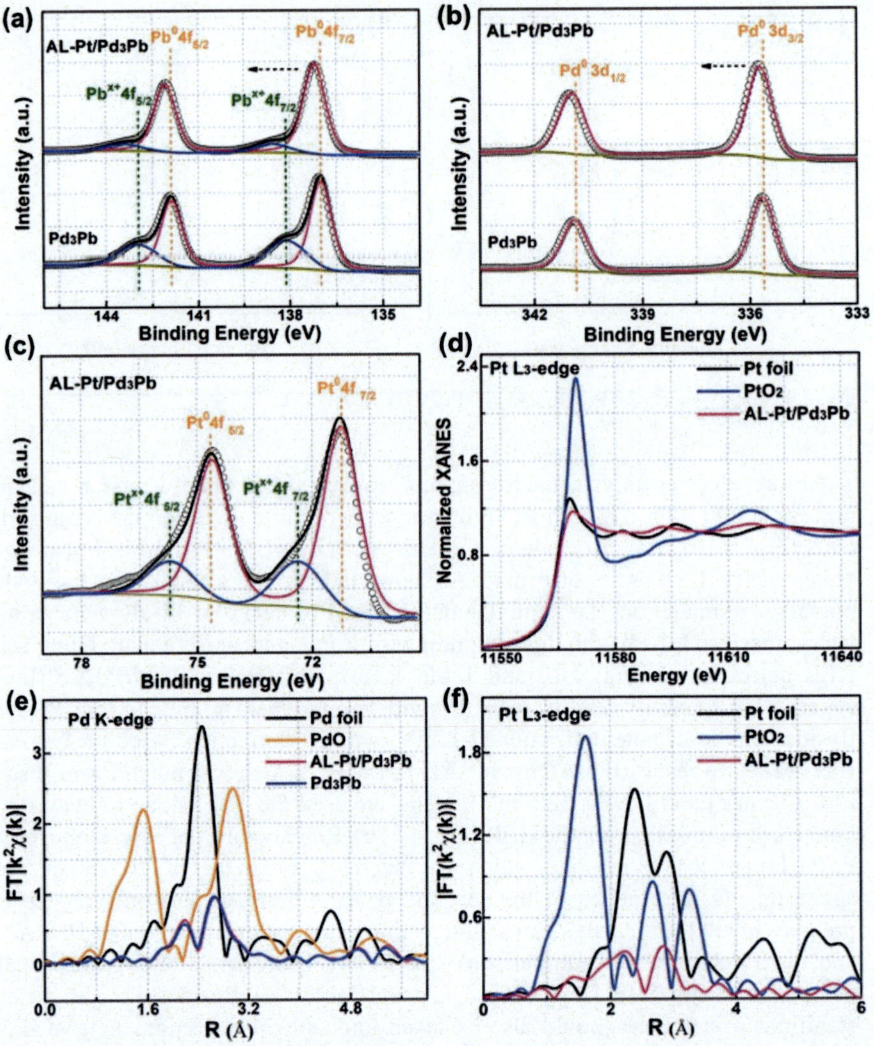

Fig. 4.9 **a–c** XPS spectra of **a** Pb 4f, **b** Pd 3d, and **c** Pt 4f. **d, e** FT-EXAFS of **d** the Pd K-edge and **e** the Pt L_3-edge. **f** XANES of the Pt L_3-edge

the Pd_3Pb substrate was also confirmed by the XRD spectra, EDS mapping, and HAADF-STEM image after the durability test (Fig. 4.23).

DFT calculations were performed to explore the origin of the high HER activity of AL-Pt/Pd_3Pb. Figure 4.24a shows a clear charge transfer from Pd and Pb to Pt, as evidenced by the electron accumulation (yellow areas) of Pt and the electron depletion (cyan areas) of Pd and Pb. The charge transfer caused an upshift of the Pt d-band center (~0.58 eV), resulting in stronger binding of H* on Pt/Pd_3Pb than on Pt(100)

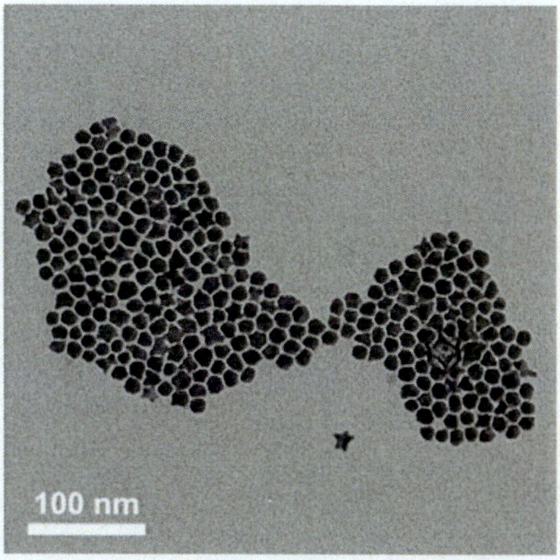

Fig. 4.10 TEM image of Pt nanoparticles

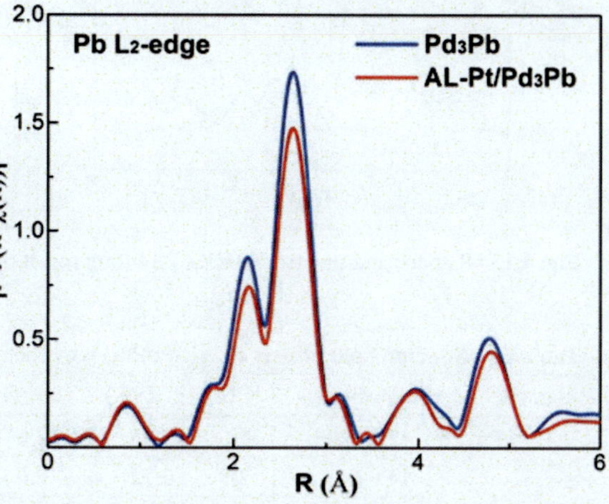

Fig. 4.11 EXAFS spectra of Pb L_2-edge

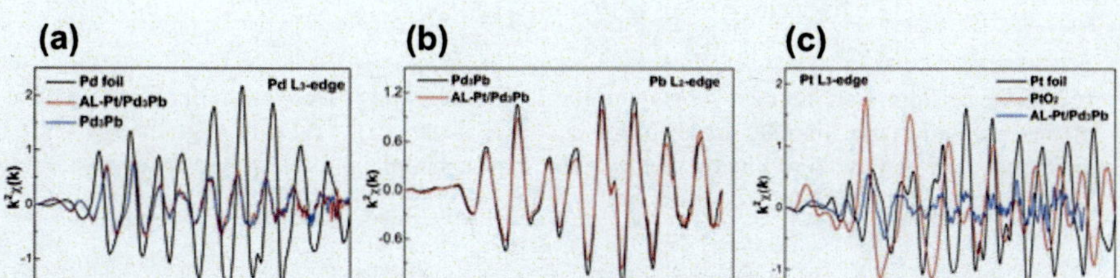

Fig. 4.12 EXAFS oscillations of **a** Pd L_3-edge, **b** Pb L_2-edge and **c** Pt L_3-edge

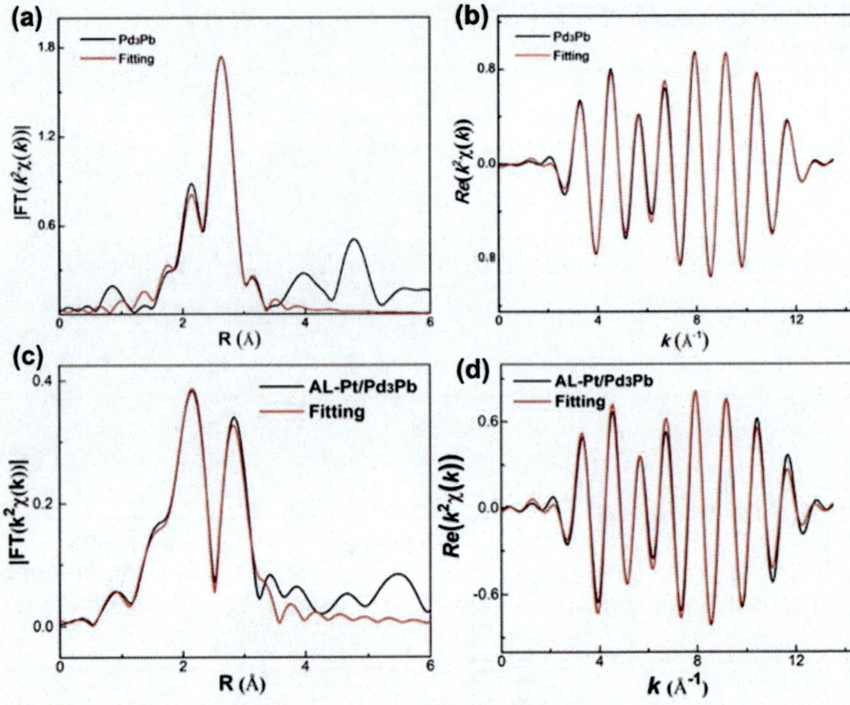

Fig. 4.13 R space and inverse FT-EXAFS fitting result of **a**, **b** Pd_3Pb and **c**, **d** AL-Pt/Pd_3Pb

Table 4.1 Structural parameters of AL-Pt/Pd_3Pb extracted from the EXAFS fitting

Sample	Path	[a]CN	[b]R (Å)	[c]σ^2 (10^{-3} Å^2)	[d]ΔE_0 (eV)	[e]R factor
AL-Pt/Pd_3Pb	Pt–Pt/Pb	5.6	2.75	12.6	1.0	0.003
	Pt–Pd	2.5	2.72	8.6	1.0	
	Pb–Pd	11.3	2.84	9.0	−1.4	0.004
	Pb–Pt	0.3	2.84	9.0	−1.4	

[a]*CN* is the coordination number; [b]*R* is interatomic distance; [c]σ^2 is Debye–Waller factor (a measure of thermal and static disorder in absorber–scatterer distances); [d]ΔE_0 is edge-energy shift (the difference between the zero kinetic energy value of the sample and that of the theoretical model); [e]*R* factor is used to value the goodness of the fitting

(Fig. 4.25). We note that the electronic structure change of Pt can also be induced by the tensile strain of Pt (~3.6%) on Pd_3Pb due to the lattice misfit between Pt and Pd_3Pb. We found that the improved H* binding on Pt was mainly from the ligand effect of Pd_3Pb (Fig. 4.25). To understand how the electronic structure change of Pt affects the HER activity, the Gibbs free energy change of H* (ΔG_{H*}) was calculated (Fig. 4.24b). We found that ΔG_{H*} was highly dependent on the coverage of H*, making the use solely of thermodynamics to evaluate the HER activity challenging. Thus, the

Table 4.2 Structural parameters of Pd_3Pb extracted from the EXAFS fitting

Sample	Path	[a]CN	[b]R (Å)	[c]σ^2 (10^{-3} Å^2)	[d]ΔE_0 (eV)	[e]R factor
Pd_3Pb	Pb–Pd	11.3	2.84	7.7	−1.6	0.001
	Pd–Pd	7.3	2.84	8.5	7.5	0.013
	Pd–Pb	3.7	2.83	8.5	0.7	

[a]*CN* is the coordination number; [b]*R* is interatomic distance (the bond length between Ru central atoms and surrounding coordination atoms); [c]σ^2 is Debye–Waller factor (a measure of thermal and static disorder in absorber–scatterer distances); [d]ΔE_0 is edge-energy shift (the difference between the zero kinetic energy value of the sample and that of the theoretical model); [e]*R* factor is used to value the goodness of the fitting

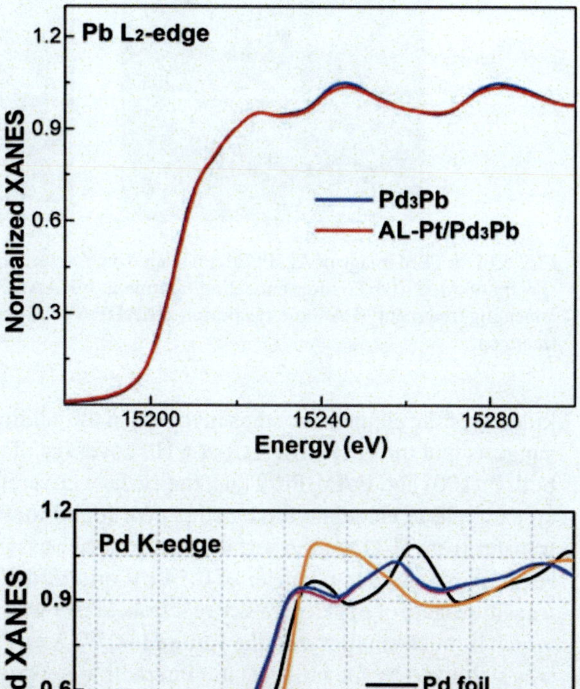

Fig. 4.14 XANES spectra of Pb L_2-edge

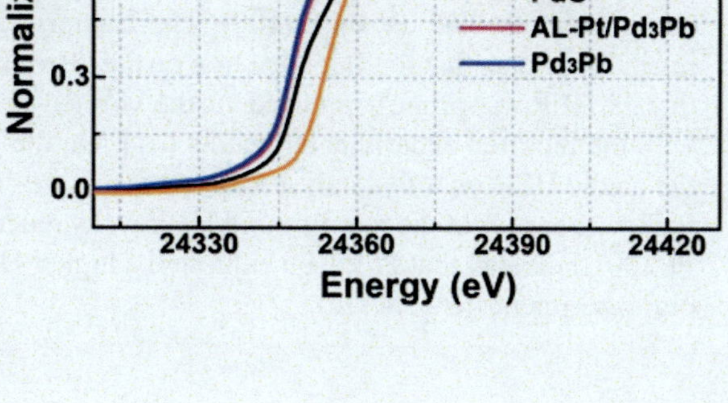

Fig. 4.15 XANES spectra of Pd K-edge

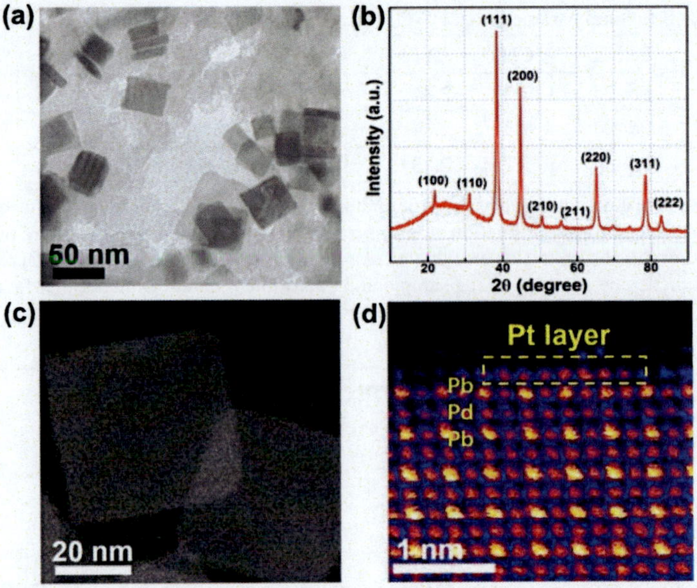

Fig. 4.16 **a** TEM image of AL-Pt/Pd_3Pb loaded on Vulcan carbon after annealing treatment. **b** XRD spectra of AL-Pt/Pd_3Pb after annealing treatment. **c** HAADF-STEM image of AL-Pt/Pd_3Pb after annealing treatment. **d** Atomic resolution HAADF-STEM image of AL-Pt/Pd_3Pb after annealing treatment

kinetics of the elementary steps involved in the acidic HER was studied. Figure 4.24b suggests that the HER will start at a H* coverage of ~1.25–1.50 monolayer (ML) on both Pt(100) and Pt/Pd_3Pb. Thus, the surface covered by 1.50 ML of H* was used as an example to calculate the barriers. We found that the Volmer reaction via proton transfer from H_3O^+ to the surface was a barrierless process on Pt(100) and was only required to pass a low barrier of 0.14 eV on Pt/Pd_3Pb (Figs. 4.24c and 4.26). At the transition state of the Heyrovsky reaction, H*/Pt was pulled out ~0.05 Å on Pt/Pd_3Pb, which is much smaller than the value of ~0.60 Å on Pt(100), since H*/Pt on Pt/Pd_3Pb was stabilized by the electrostatic interaction between the negatively charged Pt and positively charged Pd and Pb. This resulted in a lower barrier of 0.26 eV on Pt/Pd_3Pb compared with 0.62 eV on Pt(100). The barriers for the Heyrovsky reaction were much lower than that for Tafel reaction on these two surfaces (Fig. 4.24d), suggesting that the HER was prone to proceed via the Volmer–Heyrovsky mechanism, consistent with the reported experimental results [41]. On the other hand, it was reported that the acidic HER on a Pt catalyst was less structure-sensitive [42, 43], and therefore, the lower barrier of the rate-limiting Heyrovsky reaction on Pt/Pd_3Pb compared with Pt(100) indicated that Pt/Pd_3Pb exhibited a higher HER activity than Pt, in line with our experiments (Fig. 4.17a).

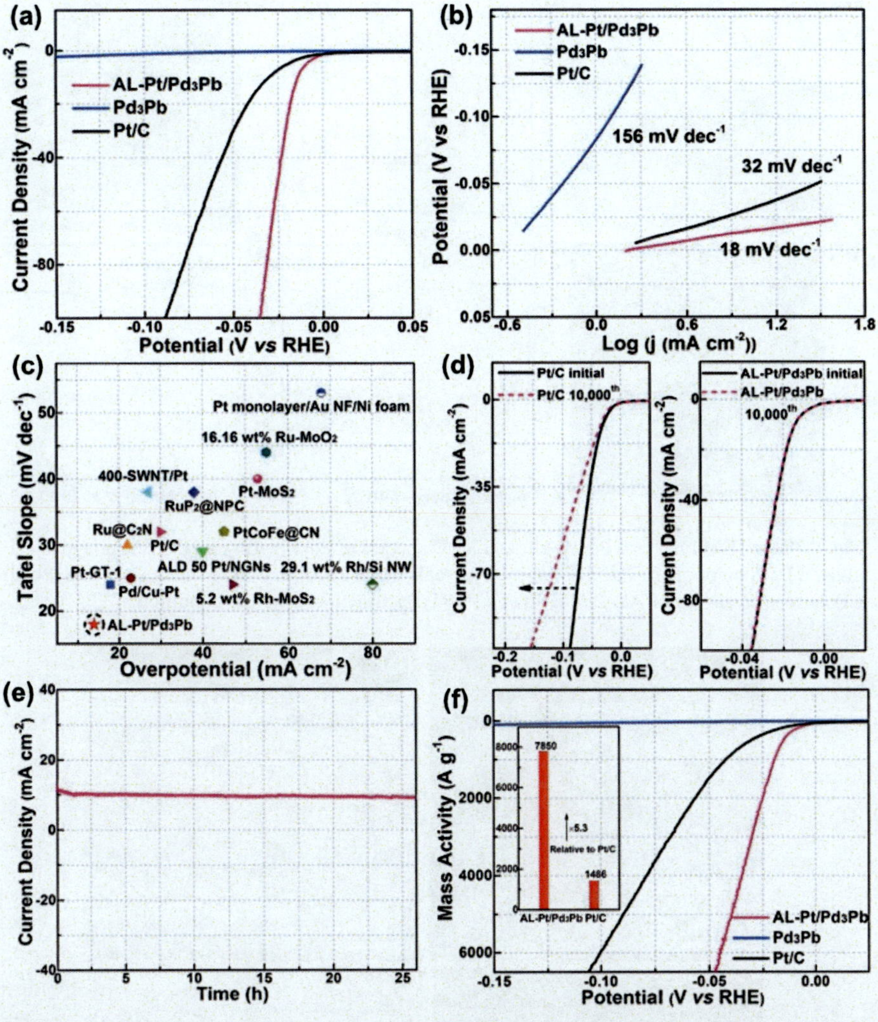

Fig. 4.17 **a** Polarization curves. **b** Tafel slopes. **c** Comparison with different representative catalysts. **d** Durability test. **e** Chronoamperometry curve for AL-Pt/Pd_3Pb. **f** Specific activity and mass activity

4.4 Conclusion

In summary, submonolayer Pt was successfully deposited on intermetallic Pd_3Pb nanoplates. The intermetallic substrate can stabilize the atomic structure of the active Pt layer and guarantee its long-term operation in the acidic HER process. The modulation of the electronic structure by the electron transfer from Pd_3Pb to Pt also results in superior HER activity of AL-Pt/Pd_3Pb. Our findings highlight the ability of surface

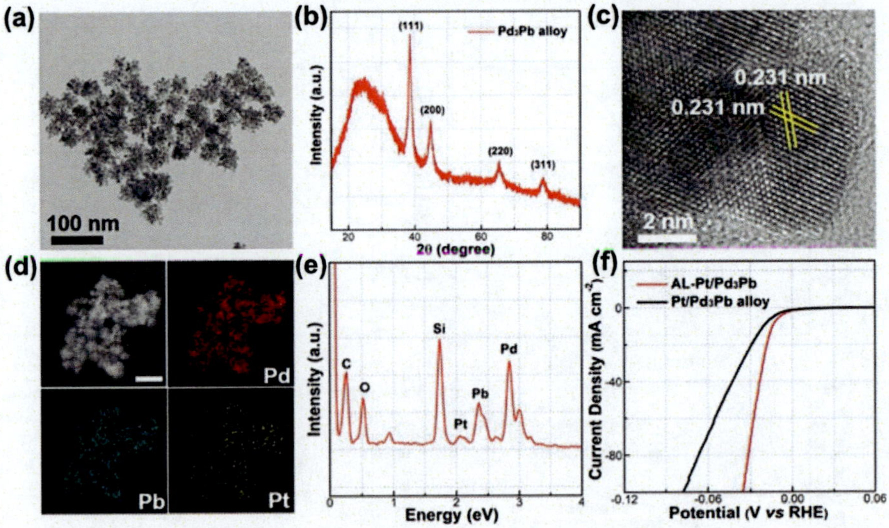

Fig. 4.18 **a** TEM image of Pd_3Pb alloy. **b** XRD spectra of Pd_3Pb alloy. **c** HRTEM image of Pd_3Pb alloy. **d** EDS mapping of Pt/Pd_3Pb alloy. The scale bar is 100 nm. **e** EDS spectra of Pt/Pd_3Pb alloy. **f** Polarization curves of AL-Pt/Pd_3Pb and Pt/Pd_3Pb alloy under acidic condition

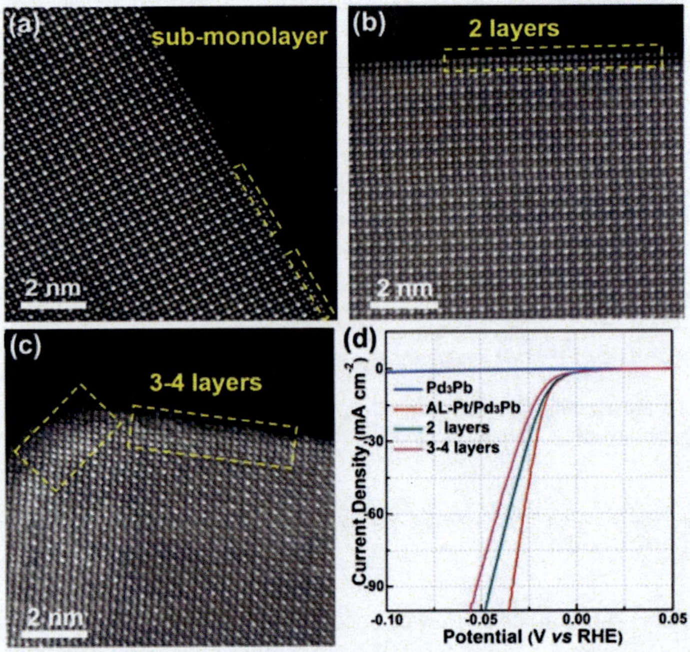

Fig. 4.19 Atomic resolution HAADF-STEM image of **a** sub-monolayer Pt on Pd_3Pb, **b** 2 layers Pt on Pd_3Pb and **c** 3–4 layers Pt on Pd_3Pb. **d** Performance comparisons of representative samples in acidic condition

Table 4.3 Comparisons of the HER activity in acidic condition

Catalysts	Overpotential at 10 mA cm^{-2} (mV)	Overpotential at 100 mA cm^{-2}
AL-Pt/Pd_3Pb	13.8	35.6
2 layers	15.3	48.1
3–4 layers	18.6	56.1

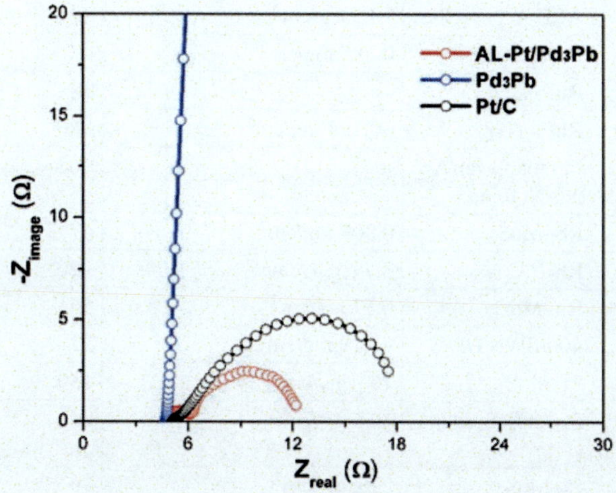

Fig. 4.20 EIS curves of AL-Pt/Pd_3Pb, Pd_3Pb and Pt/C

engineering at the atomic scale to enable extreme utilization of precious metals for energy conversion.

Table 4.4 Comparison of the activity with previously reported catalysts in acidic electrolyte

Catalyst	Catalyst loading	Overpotential at 10 mA/cm^2 (mV)	Tafel (mV/dec)	References
AL-Pt/Pd_3Pb	40.8 μg/cm^2 (1.6 μg_{Pt}/cm^2)	14	18	This work
Pt–GT-1	1.4 μg_{Pt}/cm^2	15	24	[29]
Ru@C_2N	~0.285 mg/cm^2	22	30	[30]
Pd/Cu–Pt	41 μg/cm^2	22.8	25	[15]
ALD50 Pt/NGNs	76.5 μg/cm^2	40	29	[12]
PtCoFe@CN	0.285 mg/cm^2	45	32	[31]
RuP_2@NPC	–	38	38	[32]
Rh/Si NW	~0.193 mg/cm^2	~80	24	[33]
Pt monolayer/Au NF/Ni foam	–	~68	53	[9]
Rh–MoS_2	0.309 mg/cm^2	47	24	[34]
Rh_2P/C	3.7 μg_{Rh}/cm^2	~20	–	[35]
Ru–MoO_2	0.57 mg/cm^2	55	44	[36]
400-SWNT/Pt	~0.19 mg/cm^2	27	38	[37]
Pt–MoS_2	18 μg_{Pt}/cm^2	150	96	[38]
Pt_1/mesoporous C	10 μg_{Pt}/cm^2	26	–	[39]
Pt–MoS_2	27 μg_{Pt}/cm^2	80	40	[40]

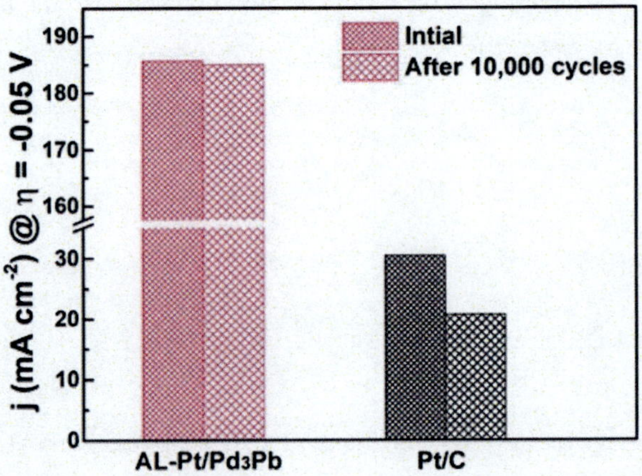

Fig. 4.21 Specific activity at −0.05 V of AL-Pt/Pd_3Pb and commercial Pt/C for the HER before and after 10,000 cycles

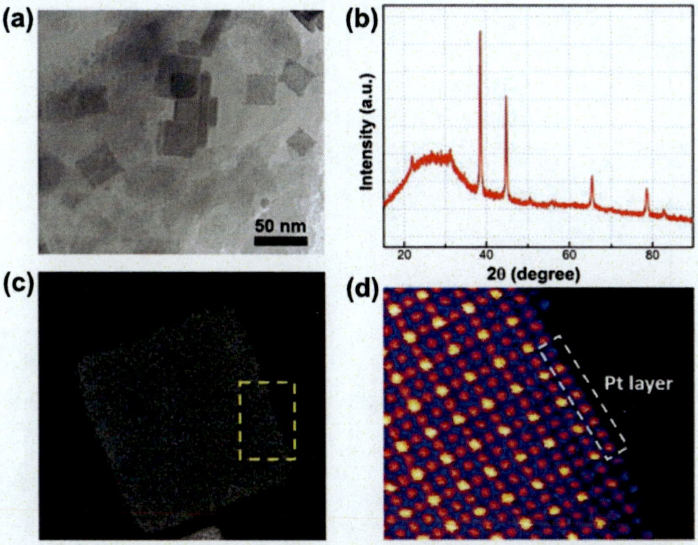

Fig. 4.22 **a** TEM image of AL-Pt/Pd_3Pb loaded on Vulcan carbon after HER durability test. **b** XRD spectra of AL-Pt/Pd_3Pb after HER durability test. **c** HAADF-STEM image of AL-Pt/Pd_3Pb nanoplate after HER durability test. **d** Atomic resolution HAADF-STEM image of an individual nanoplate at selected area (as shown in yellow dotted box in Fig. S17c)

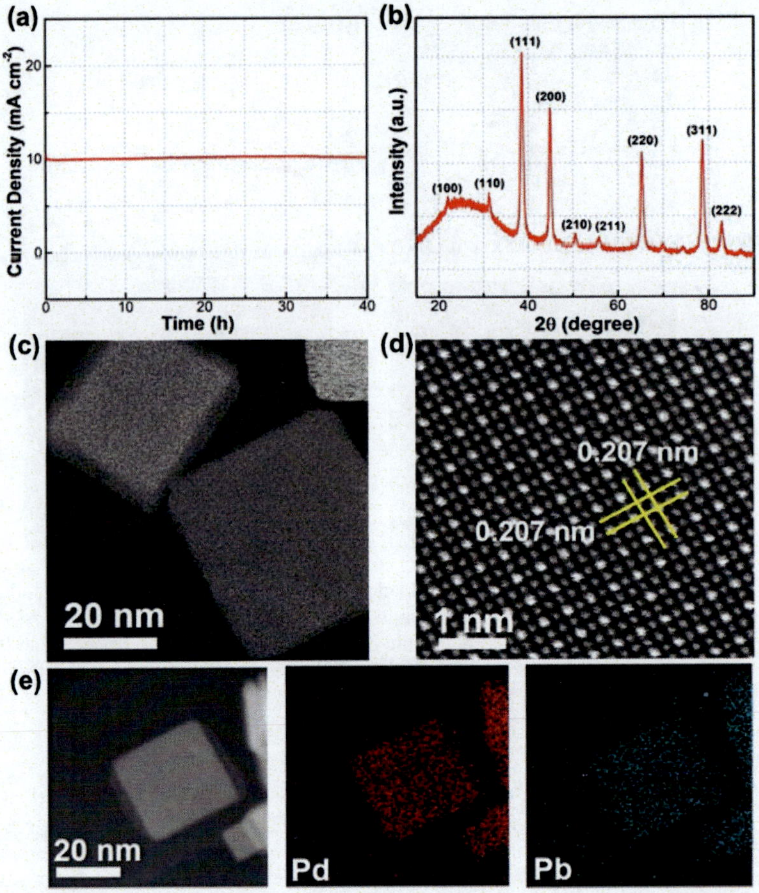

Fig. 4.23 **a** Chronoamperometry curve for Pd_3Pb at a potential of −0.3 V. **b** XRD spectra of Pd_3Pb after durability test. **c** HAADF-STEM image of Pd_3Pb after durability test. **d** Atomic resolution HAADF-STEM image after durability test. **e** EDS mapping of Pd_3Pb after durability test

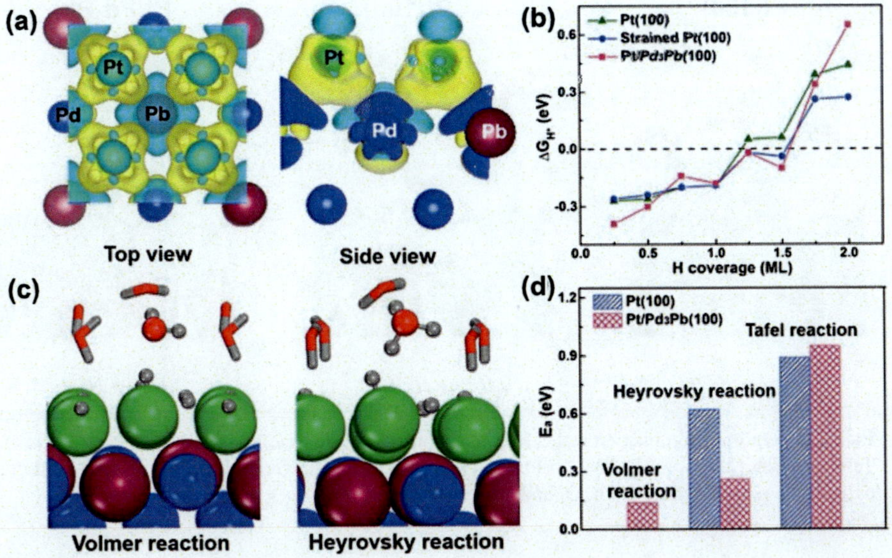

Fig. 4.24 **a** Charge density difference (0.003 e/Å^3) of Pt and Pd_3Pb in Pt/Pd_3Pb(100). **b** Calculated differential free energy change of H*. **c** Optimized transition states for the Volmer and Heyrovsky reactions on Pt/Pd_3Pb. **d** Calculated barriers

Fig. 4.25 Calculated average binding energy of H* at various hydrogen coverages on Pt(100), strained Pt(100), and Pt/Pd_3Pb(100)

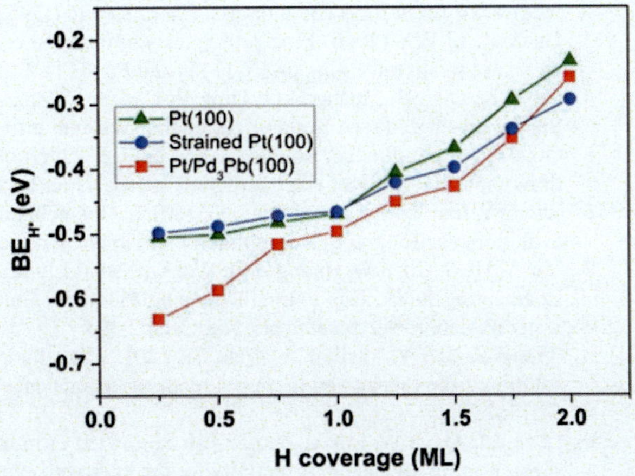

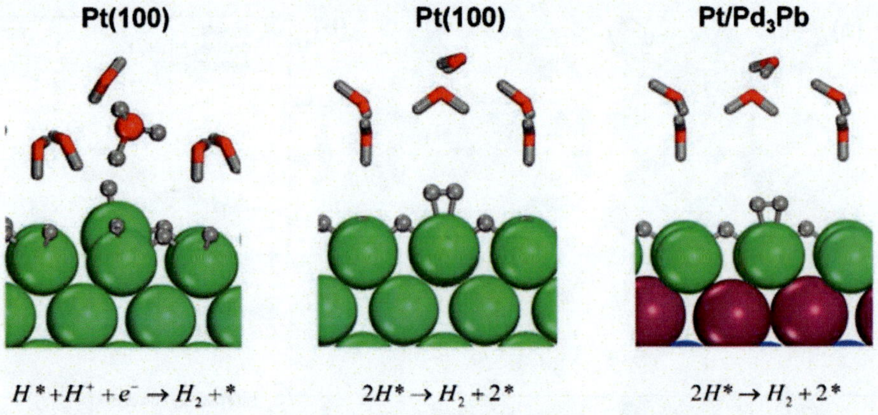

Fig. 4.26 Optimized transition states for the Heyrovsky reaction (H $*$ $+H^{+}+e^{-} \rightarrow H_2 + *$) and Tafel reaction ($2H* \rightarrow H_2 + *$) on Pt(100) and Pt/$Pd_3Pb$. The green, blue, purple, red, and gray balls/sticks represent Pt, Pd, Pb, O, and H atoms, respectively

References

1. Palo DR, Dagle RA, Holladay JD (2007) Methanol steam reforming for hydrogen production. Chem Rev 107:3992
2. Navarro RM, Pena MA, Fierro JLG (2007) Hydrogen production reactions from carbon feedstocks: fossil fuels and biomass. Chem Rev 107:3952
3. Gu X-K, Li WX (2010) First-principles study on the origin of the different selectivities for methanol steam reforming on Cu (111) and Pd (111). J Phys Chem C 114:21539
4. Gu X-K, Qiao B, Huang C-Q, Ding W-C, Sun K, Zhan E, Zhang T, Liu J, Li W-X (2014) Supported single Pt_1/Au_1 atoms for methanol steam reforming. ACS Catal 4:3886
5. Greeley J, Jaramillo TF, Bonde J, Chorkendorff IB, Nørskov JK (2006) Computational high-throughput screening of electrocatalytic materials for hydrogen evolution. Nat Mater 5:909
6. Zheng Y, Jiao Y, Jaroniec M, Qiao SZ (2015) Advancing the electrochemistry of the hydrogen-evolution reaction through combining experiment and theory. Angew Chem Int Ed 54:52
7. Yao Y, Hu S, Chen W, Huang Z-Q, Wei W, Yao T, Liu R, Zang K, Wang X, Wu G et al (2019) Engineering the electronic structure of single atom Ru sites via compressive strain boosts acidic water oxidation electrocatalysis. Nat Catal 2:304
8. Zheng Y, Jiao Y, Vasileff A, Qiao SZ (2018) The hydrogen evolution reaction in alkaline solution: from theory, single crystal models, to practical electrocatalysts. Angew Chem Int Ed 57:7568
9. Li M, Ma Q, Zi W, Liu X, Zhu X, Liu SF (2015) Pt monolayer coating on complex network substrate with high catalytic activity for the hydrogen evolution reaction. Sci Adv 1:e1400268
10. Morales-Guio CG, Stern LA, Hu X (2014) Nanostructured hydrotreating catalysts for electrochemical hydrogen evolution. Chem Soc Rev 43:6555
11. Subbaraman R, Tripkovic D, Strmcnik D, Chang K-C, Uchimura M, Paulikas AP, Stamenkovic V, Markovic NM (2011) Enhancing hydrogen evolution activity in water splitting by tailoring Li^{+}-$Ni(OH)_2$-Pt interfaces. Science 334:1256
12. Cheng N, Stambula S, Wang D, Banis MN, Liu J, Riese A, Xiao B, Li R, Sham T-K, Liu L-M, Botton GA, Sun X (2016) Platinum single-atom and cluster catalysis of the hydrogen evolution reaction. Nat Commun 7:13638
13. Yu W, Porosoff MD, Chen JG (2012) Review of Pt-based bimetallic catalysis: from model surfaces to supported catalysts. Chem Rev 112:5780

14. Stamenkovic VR, Fowler B, Mun BS, Wang G, Ross PN, Lucas CA, Markovic NM (2007) Improved oxygen reduction activity on Pt_3Ni (111) via increased surface site availability. Science 315:493
15. Chao T, Luo X, Chen W, Jiang B, Ge J, Lin Y, Wu G, Wang X, Hu Y, Zhuang Z, Wu Y, Hong X, Li Y (2017) Atomically dispersed copper-platinum dual sites alloyed with palladium nanorings catalyze the hydrogen evolution reaction. Angew Chem Int Ed 56:16047
16. Zhang J, Zhao Y, Guo X, Chen C, Dong C-L, Liu R-S, Han C-P, Li Y, Gogotsi Y, Wang G (2018) Single platinum atoms immobilized on an MXene as an efficient catalyst for the hydrogen evolution reaction. Nat Catal 1:985
17. Liu Y, Gokcen D, Bertocci U, Moffat TP (2012) Self-terminating growth of platinum films by electrochemical deposition. Science 338:1327
18. Wang X, Choi S-I, Roling LT, Luo M, Ma C, Zhang L, Chi M, Liu J, Xie Z, Herron JA, Mavrikakis M, Xia Y (2015) Palladium-platinum core-shell icosahedra with substantially enhanced activity and durability towards oxygen reduction. Nat Commun 6:7594
19. Meier JC, Galeano C, Katsounaros I, Topalov AA, Kostka A, Schüth F, Mayrhofer KJ (2012) Degradation mechanisms of Pt/C fuel cell catalysts under simulated start-stop conditions. ACS Catal 2:832
20. Yang XF, Wang A, Qiao B, Li J, Liu J, Zhang T (2013) Single atom catalysts: a new frontier in heterogeneous catalysis. Acc Chem Res 46:1740
21. Pallassana V, Neurock M, Hansen LB, Hammer B, Nørskov JK (1999) Theoretical analysis of hydrogen chemisorption on Pd(111), Re(0001) and Pd_{ML}/Re(0001), Re_{ML}/Pd(111) paedomorphic overlayers. Phys Rev B Condens Matter Mater Phys 60:6146
22. Seh ZW, Kibsgaard J, Dickens CF, Chorkendorff IB, Nørskov JK, Jaramillo TF (2017) Combining theory and experiment in electrocatalysis: insights into materials design. Science 355:eaad4998
23. Casado-Rivera E, Volpe DJ, Alden L, Lind C, Downie C, Vazquez Alvarez T, Angelo AC, DiSalvo FJ, Abruna HD (2004)´ Electrocatalytic activity of ordered intermetallic phases for fuel cell applications. J Am Chem Soc 126:4043
24. Yan Y, Du JS, Gilroy KD, Yang D, Xia Y, Zhang H (2017) Intermetallic nanocrystals: syntheses and catalytic applications. Adv Mater 29:1605997
25. Li J, Yin H-M, Li X-B, Okunishi E, Shen Y-L, He J, Tang Z-K, Wang W-X, Yücelen E, Li C et al (2017) Surface evolution of a Pt–Pd–Au electrocatalyst for stable oxygen reduction. Nat Energy 2:17111
26. Kothleitner G, Neish MJ, Lugg NR, Findlay S, Grogger W, Hofer F, Allen LJ (2014) Quantitative elemental mapping at atomic resolution using X-ray spectroscopy. Phys Rev Lett 112:085501
27. Choudhury D, Das B, Sarma DD, Rao CNR (2010) XPS evidence for molecular charge-transfer doping of graphene. Chem Phys Lett 497:66
28. Chen C, Kang Y, Huo Z, Zhu Z, Huang W, Xin HL, Snyder JD, Li D, Herron JA, Mavrikakis M, Chi M, More KL, Li Y, Markovic NM, Somorjai GA, Yang P, Stamenkovic VR (2014) Highly crystalline multimetallic nanoframes with three-dimensional electrocatalytic surfaces. Science 343:1339
29. Tiwari JN, Sultan S, Myung CW, Yoon T, Li N, Ha M, Harzandi AM, Park HJ, Kim DY, Chandrasekaran SS (2018) Multicomponent electrocatalyst with ultralow Pt loading and high hydrogen evolution activity. Nat Energy 3:773
30. Mahmood J, Li F, Jung S-M, Okyay MS, Ahmad I, Kim S-J, Park N, Jeong HY, Baek J-B (2017) An efficient and pH-universal ruthenium-based catalyst for the hydrogen evolution reaction. Nat Nanotechnol 12:441
31. Chen J, Yang Y, Su J, Jiang P, Xia G, Chen Q (2017) Enhanced activity for hydrogen evolution reaction over CoFe catalysts by alloying with small amount of Pt. ACS Appl Mater Interfaces 9:3596
32. Pu Z, Amiinu IS, Kou Z, Li W, Mu S (2017) RuP_2-based catalysts with platinum-like activity and higher durability for the hydrogen evolution reaction at all pH values. Angew Chem 129:11717

33. Zhu L, Lin H, Li Y, Liao F, Lifshitz Y, Sheng M, Lee S-T, Shao M (2016) A rhodium/silicon co-electrocatalyst design concept to surpass platinum hydrogen evolution activity at high overpotentials. Nat Commun 7:12272
34. Cheng Y, Lu S, Liao F, Liu L, Li Y, Shao M (2017) Rh-MoS_2 nanocomposite catalysts with Pt-like activity for hydrogen evolution reaction. Adv Funct Mater 27:1700359
35. Duan H, Li D, Tang Y, He Y, Ji S, Wang R, Lv H, Lopes PP, Paulikas AP, Li H (2017) High-performance Rh_2P electrocatalyst for efficient water splitting. J Am Chem Soc 139:5494
36. Jiang P, Yang Y, Shi R, Xia G, Chen J, Su J, Chen Q (2017) Pt-like electrocatalytic behavior of Ru-MoO_2 nanocomposites for the hydrogen evolution reaction. J Mater Chem A 5:5475
37. Tavakkoli M, Holmberg N, Kronberg R, Jiang H, Sainio J, Kauppinen EI, Kallio T, Laasonen K (2017) Electrochemical activation of single-walled carbon nanotubes with pseudo-atomic-scale platinum for the hydrogen evolution reaction. ACS Catal 7:3121
38. Deng J, Li H, Xiao J, Tu Y, Deng D, Yang H, Tian H, Li J, Ren P, Bao X (2015) Triggering the electrocatalytic hydrogen evolution activity of the inert two-dimensional MoS_2 surface via single-atom metal doping. Energy Environ Sci 8:1594
39. Dendooven J, Ramachandran RK, Solano E, Kurttepeli M, Geerts L, Heremans G, Rongé J, Minjauw MM, Dobbelaere T, Devloo-Casier K (2017) Independent tuning of size and coverage of supported Pt nanoparticles using atomic layer deposition. Nat Commun 8:1074
40. Huang X, Zeng Z, Bao S, Wang M, Qi X, Fan Z, Zhang H (2013) Solution-phase epitaxial growth of noble metal nanostructures on dispersible single-layer molybdenum disulfide nanosheets. Nat Commun 4:1444
41. Markovic NM, Grgur BN, Ross PN (1997) Temperature dependent hydrogen electrochemistry on platinum low-index single crystal surfaces in acid solutions. J Phys Chem B 101:5405
42. Kita H, Ye S, Gao Y (1992) Mass transfer effect in hydrogen evolution reaction on Pt single-crystal electrodes in acid solution. J Electroanal Chem 334:351
43. Gomez R, Fernandez-Vega A, Feliu JM, Aldaz AJ (1993) Hydrogen evolution on Pt single crystal surfaces: effects of irreversibly adsorbed bismuth and antimony on hydrogen adsorption and evolution on Pt(100). J Phys Chem C 97:4769

Chapter 5
Conclusion and Perspective

Noble metal nanocrystals that are regarded as promising candidates have been widely applied in energy conversion and storage. This thesis is mainly involved in the controllable synthesis of novel ternary noble metal nanocrystals and their applications, of which are briefly introduced in the following.

We report an epitaxial-growth-mediated method to grow face-centered cubic (fcc) Ru, which is thermodynamically unfavorable in the bulk form, on the surface of Pd–Cu alloy. Induced by the galvanic replacement between Ru and Pd–Cu alloy, a shape transformation from a Pd–Cu@Ru core–shell to a yolk–shell structure was observed during the epitaxial growth. The successful coating of the unconventional crystallographic structure is critically dependent on the moderate lattice mismatch between the fcc Ru overlayer and $PdCu_3$ alloy substrate. Further, both fcc and hexagonal close packed (hcp) Ru can be selectively grown through varying the lattice spacing of the Pd–Cu substrate. The presented findings provide a new synthetic pathway to control the crystallographic structure of metal nanomaterials.

We have synthesized a series of $PtCu_x/Pt_{skin}$ core–shell structures with atomically dispersed Ru_1 and have unravelled their mechanism of formation, oxidation resistance and the origin of the enhanced OER catalysis by exhaustively examining their structures, coordination environments and oxidation states. Through sequential acid etching and electrochemical leaching, the structure of $PtCu_x$ alloys can be varied (to give $PtCu_3$, PtCu and Pt_3Cu), which modulates effectively the OER activity catalysed by the Ru_1. The best catalyst, Ru_1–Pt_3Cu, delivers 220 mV overpotential to achieve a current density of 10 mA cm^{-2} for acidic OER, with ten times longer lifetime over commercial RuO_2. We found that there is a volcano-type relation between the OER activity and the lattice constant. We argue that the compressive strain of the Pt_{skin} shell effectively engineers the electronic structure and redox behaviour of single atomic Ru anchored at the corner or step sites of the Pt-rich shell, with optimized binding of oxygen intermediates and better resistance to over-oxidation and dissolution.

Submonolayer Pt was controllably deposited on an intermetallic Pd_3Pb nanoplate (AL-Pt/Pd_3Pb). The atomic efficiency and electronic structure of the active surface Pt layer were largely optimized, greatly enhancing the acidic HER. AL-Pt/Pd_3Pb

Y. Yao, *Controllable Synthesis and Atomic Scale Regulation of Noble Metal Catalysts*, Springer Theses, https://doi.org/10.1007/978-981-19-0205-5_5

exhibits an outstanding HER activity with an overpotential of only 13.8 mV at 10 mA/cm^2 and a high mass activity of 7834 A/g_{Pd+Pt} at −0.05 V, both largely surpassing those of commercial Pt/C (30 mV, 1486 A/g_{Pt}). In addition, AL-Pt/Pd_3Pb shows excellent stability and robustness. Theoretical calculations show that the improved activity is mainly derived from the charge transfer from Pd_3Pb to Pt, resulting in a strong electrostatic interaction that can stabilize the transition state and lower the barrier.

Our work not only provides new strategies to design efficient noble metal nanomaterials, but also highlights the underlying catalytic mechanism in electrochemistry, such as OER, HER. In the future, we would devote to develop in-situ characterizing technologies (e.g., XAFS, FTIR) to understand catalytic mechanism at the molecule level aiming at controlled preparation of higher-performing nanomaterials.

Author Biography

Dr. Yancai Yao received her doctoral degree from The University of Science and Technology of China (USTC). Her research interests are focused on controlled preparation of noble metal nanomaterials and energy conversion/storage application.

Y. Yao, *Controllable Synthesis and Atomic Scale Regulation of Noble Metal Catalysts*, Springer Theses, https://doi.org/10.1007/978-981-19-0205-5